国粹赓续卷——历史建筑保护与再利用设计

National Quintessence Continuation—Historic Building Conservation and Reuse Design

中国建筑学会室内设计分会推荐专业教学参考书

"室内设计 6+1" 2018（第六届）联合毕业设计

中国建筑学会室内设计分会
北京建筑大学 编

·北京·

图书在版编目（CIP）数据

国粹赓续卷：历史建筑保护与再利用设计：中国建筑学会室内设计分会推荐专业教学参考书 "室内设计6+" 2018（第六届）联合毕业设计 / 中国建筑学会室内设计分会，北京建筑大学编. -- 北京：中国水利水电出版社，2018.10
　ISBN 978-7-5170-7037-5

Ⅰ. ①国… Ⅱ. ①中… ②北… Ⅲ. ①古建筑—文物保护—设计方案—作品集—中国—现代 Ⅳ. ①TU-87

中国版本图书馆CIP数据核字(2018)第239228号

书　　名	中国建筑学会室内设计分会推荐专业教学参考书 "室内设计6+" 2018（第六届）联合毕业设计 **国粹赓续卷——历史建筑保护与再利用设计** GUOCUI GENGXU JUAN—LISHI JIANZHU BAOHU YU ZAILIYONG SHEJI
作　　者	中国建筑学会室内设计分会　北京建筑大学　编
出版发行	中国水利水电出版社 （北京市海淀区玉渊潭南路1号D座　100038） 网址：www.waterpub.com.cn E-mail：sales@waterpub.com.cn 电话：(010) 68367658（营销中心）
经　　售	北京科水图书销售中心（零售） 电话：(010) 88383994、63202643、68545874 全国各地新华书店和相关出版物销售网点
排　　版	中国建筑学会室内设计分会
印　　刷	北京印匠彩色印刷有限公司
规　　格	210mm×285mm　16开本　16.5印张　855千字
版　　次	2018年10月第1版　2018年10月第1次印刷
印　　数	0001—1000册
定　　价	120.00元

凡购买我社图书，如有缺页、倒页、脱页的，本社营销中心负责调换
版权所有·侵权必究

编委会
Editorial Committee

Director 主任
Su Dan, Li Aiqun　苏　丹　李爱群

Deputy Director 副主任（以室内分会和参编院校为序）
Ye Hong, Li Junqi, Li Zhenyu, Zhang Ke, Sun Cheng　叶　红　李俊奇　李振宇　张　珂　孙　澄
Lin Baogang, Ouyang Wen, Li Yiwen, Chen Qianhu　蔺宝钢　欧阳文　李亦文　陈前虎

Judges and Corporate Mentor 评委及企业导师

Chief Judge 本届评委组长
Song Weijian　宋微建

Former Chief Judge 往届评委组长
Zhao Jian, Zhuo Pei, Chen Weixin,　赵　健（第五届）　卓　培（第四届）　陈卫新（第三届）
Wang Chuanshun, Guo Xiaoming　王传顺（第二届）　郭晓明（第一届）

Committee member 委员（以姓名拼音字母为序）
Cui Lin, Huang Quan, Kou Jianchao, Li Zhengtao, Liu Lei, Lou Chunyu,　崔　林　黄　全　寇建超　李正涛　刘　磊　娄春雨
Sun Dongning, Wang Chuanshun, Wang Ye, Wang Weimin, Xing Ye, Yao Li,　孙冬宁　王传顺　王　野　王炜民　幸　晔　姚　丽
Yao Ling, Ye Zheng, Zeng Jun, Zhu Haixuan　姚　领　叶　铮　曾　军　朱海玄

In-school tutor 高校导师（以姓名拼音字母为序）
Lin yi, Liu Xiaojun, Lv Qinzhi, Ma hui, Shi Tuo, Wang Chen　林　怡　刘晓军　吕勤智　马　辉　石　拓　王　琛
Wang Min, Wang Yihan, Xie Guanyi, Yang Lin, Zhao Hui, Zhou Lijun　王　敏　王一涵　谢冠一　杨　琳　兆　翚　周立军
Zhu Fei, Zuo Yan　朱　飞　左　琰

Chief Editor 执行主编
Chen Jingyong　陈静勇

Associate Editor 执行副主编
Zhang Kefan, Pan Xiaowei　张可凡　潘晓微

Editorial Committee 执行编委
Sun Xiaopeng, Wang Yan　孙小鹏　王艳

Book Design 书籍设计
Zhang Kefan, Wang Yan　张可凡　王艳

Cover Design 封面设计
Wang Yan　王艳

北京建筑大学设计艺术研究院
BUCEA ACADEMY OF DESIGN AND ART

出版支持：北京建筑大学教材建设专项

内容提要

历史建筑是指经城市、县人民政府确定公布的具有一定保护价值，能够反映历史风貌和地方特色的建筑物、构筑物，是城市发展演变历程中留存下来的重要历史载体。近年来，对历史建筑的保护与再利用问题研究与实践已成为我国快速城市化发展进程中的焦点和热点之一。加强历史建筑的保护和合理利用，有利于展示城市历史风貌，保留城市的建筑风格和文化特色，是践行新发展理念、树立文化自信的一项重要工作。为此，中国建筑学会室内设计分会将「国粹赓续——历史建筑保护与再利用设计」作为「室内设计6+」2018（第六届）联合毕业设计的总命题。

中国「室内设计6+」联合毕业设计是中国建筑学会室内设计分会2013年创立的具有特色的设计教育创新项目。同济大学、华南理工大学、哈尔滨工业大学、西安建筑科技大学、北京建筑大学、南京艺术学院、浙江工业大学为本届活动的参加高校，分别联合企业共同指导毕业设计。来自建筑学、环境设计、工业设计、艺术与科技等专业2018届的毕业生们，通过与学会、高校、企业、专家间的协同合作，就「国粹赓续——历史建筑保护与再利用设计」总命题下的子课题，联合开展了综合性实践教学活动，对多维视野下的历史建筑及其价值、历史建筑保护和再利用的历史与理念、历史建筑再利用及其新旧关联模式、新旧空间关联及其相关建筑理论等成为多专业领域的共同关注点等方面进行探讨和深入交流，促进了毕业设计教学水平和人才培养质量的提升。

据此，中国建筑学会室内设计分会组织项目的参与高校共同编著了中国建筑学会室内设计分会推荐专业教学参考书：「室内设计6+」联合毕业设计《国粹赓续卷——历史建筑保护与再利用设计》。全书分为项目规章、调研踏勘、中期检查、答辩展示、专家讲坛、教育研究、风采定格等特色章节，主要内容采用中英文对照方式，记载了本届设计教育创新项目的开展印记，图文并茂，内容详实。

本书可供建筑学、环境设计、工业设计、艺术与科技，以及城乡规划、风景园林、城市地下空间工程、产品设计、视觉传达设计、公共艺术等专业人员及设置相关专业的院校师生参考借鉴。

序
Preface

苏丹
中国建筑学会室内设计分会 理事长
President of the China Institute of Interior Design
清华大学美术学院 教授
A professor at School of Fine Arts, Tsinghua University

出师表
Memorial to the Throne for an Expedition

看到了持续六年的"室内设计6+"联合毕业设计特色教育创新项目以及其累计的丰硕成果，我突然想到了《出师表》。也许有人会说相隔了一千八百年的时光，又是两种类型的事情，一个是出征前的资政，一个是学术活动的总结。它们之间有什么联系呢？我说：当然有！

首先触动我的，是我们这个学术组织发展的历程和初心。1989年中国建筑学会室内设计分会的前身中国室内建筑师学会成立，在当时这是一件非常具有远见卓识的大事情，事实也如那些开疆拓土的前辈们预料到一样，室内设计这个学科领域、这个行业、这项事业在过去的29年中已是江河万里、沧海桑田。用巨变来描述毫不为过，而更加重要的是质变，即"室内设计"逐渐地完善也是一个基本事实。数以百万计的室内设计师，服务着数以亿计的人民大众；每年数以百亿的设计产值，同时撬动着数以万亿计的国民经济产值；此外，室内设计正在迅速地提高人民的生活水平，不断拉近我们和文明的距离。在学术层面，一方面，它的工程属性得到了重视和发展，这得益于几十年来学会所推动的建筑系统性；另一方面，它的文化和艺术属性不断加强，这也得益于学会倡导的包容性。

然而在一个逐利的时代，面对这样的一个潜力无穷又商机无限的领域，不同的群体从各自的出发点假以各式各样的口号发起不同的组织。这些组织如飘浮的冰层，在洋流的作用下，在同一领域经常彼此摩擦、相互撞击。另一方面，同样的一群设计师经常拥有多个身份，在不同的浮冰之间来回穿梭。这种群雄并起逐鹿天下的局面，总令我想起《出师表》开篇的那些对天下格局的概述"今天下三分，益州疲敝，此诚危急存亡之秋也"。

近几年这种忧虑越来越重，同质化的竞争在多个方面严重影响着这个领域的健康发展。

《出师表》第二段"然侍卫之臣不懈于内，忠志之士忘身于外者，盖追……"，这句话谈到了"坚守者"。这是让人非常感动的地方，你会觉得在这个纷乱的时代依然

I thought of Memorial to the Throne for an Expedition at the sight of the Innovation Program of Characteristic Education on "interior Design 6+" Joint Graduation Project and its fruitful achievements. It can be argued that they are 1800 years apart and two types of things, one being a memorial to the throne before going into battle, the other being a summary of an academic activity, so what's the connection between them? I would answer, "Of course there is a connection between them!"

The first thing to touch me is the development history and original aspiration of the academic organization. The China Institute of Interior Design (CIID), formerly known as the Society of Chinese Interior Architects, was founded in 1989 with farsightedness and insight. As expected by our predecessors that broke the ground, as a discipline, an industry and an undertaking, interior design has achieved great development and great results in the past 29 years. Its development can be described as a radical change, and what matters the most is the qualitative change, in other words, it is also a basic fact that "Interior design" has been gradually perfected in essence. Millions of interior designers serve billions of common people; interior design has an annual output value of tens of billions of yuan, and makes an indirect contribution to multi-trillion-yuan national economic output; besides, interior design is rapidly improving the people's living standards and shortening the distance between civilization and us. At the academic level, on the one hand, its engineering attributes have been valued and developed. This benefits from the architectural systematization that has been promoted by the ASC in the past several decades; on the other hand, its cultural and artistic attributes are being strengthened. This owes to the inclusiveness espoused by the ASC.

However, in this money-oriented era, faced with such a field with infinite potential and unlimited business opportunities, different groups have founded different organizations with different slogans from their own perspectives. These organizations, like floating ices, often rub and collide against one another in the same field under the action of ocean currents. On the other hand, a designer may have many identities and shuttle back and forth among different floating ices. The contention among different design schools reminds me of the introductory overview of the world situation in Memorial to the Throne for an Expedition: "Today, the empire remains divided in three, and our very survival is threatened".

In recent years, this has been increasingly worrisome for us, and homogeneity competition has seriously affected the sound development of this field in many ways.

The second paragraph of Memorial to the Throne for an Expedition says, "Yet still the officials at court and the soldiers throughout the realm remain loyal to you, your majesty. Because they remember the late emperor…". This sentence mentions "the defender". It reads moving. You may realize that in this era of chaos, there are still people holding fast to the academic value, doing research independently, and working on sacred education. The

有一群人坚守学术的价值，去做孤独的研究、做神圣的教育。"室内设计6+"联合毕业设计特色教育创新项目就是这种类型的一种鲜明代表，这个项目的主导者就是时代的坚守者。长期以来中国建筑学会室内设计分会关注设计教育，而这个领域竟成为一个行业热点中的高海拔地区。高处不胜寒，以此见真心。六年以来，项目核心高校室内设计、环境设计等学科骨干密切协同，他们把学科建设与社会服务通过对现实问题的回应结合了起来。采取了生动活泼的校内校外、课上课下的教学和研究，可谓硕果累累。我为这些同行勤勉的工作感到骄傲！

在当下的室内设计领域，一个突出实用价值的阶段，唯有高等教育机构还在担当着研究和思考的重任。无论是基础研究还是对现实个案展开的深层次研究，都是在学科里进行的，这不仅弥足珍贵，更不可或缺。作为一个以学术研究为己任的社会团体，这种性质的工作才是我们的本色，因此我们必须对他们长期以来的努力给予支持和帮助。

《出师表》在行事的方法上有许多精辟的概述，如"……悉以咨之，然后施行，必能裨补阙漏，有所广益。"在室内设计学科领域，研究和实践在时间上的关系也始终处于螺旋上升的结构状态，实践是研究的开始，研究又是大规模实践的准备。我一直认为室内设计有着广阔的未来，它将是延续几千年的建筑设计未来的存在形式，需要"咨诹善道，察纳雅言"对其本体进行更多的分析研究，对它的未来展开更多的讨论和展望。

苏丹
2018年9月 于米兰

Innovation Program of Characteristic Education on "interior Design 6+" Joint Graduation Project is a distinct representative of this type, and the dominator of this program is the defender of the era. For a long time, the IID-ASC has been concerned about design education, which has, unexpectedly, become a highland of the industrial hotspots. It's lonely to be in a high position, but sincerity can be seen this way. Over the six years, the backbones in interior design and environment design at the universities involved have worked closely with one another, combining discipline construction with social services by responding to practical problems. The have done lively teaching and research work on and off campus, in and outside class, making innumerable great achievements. I'm proud of their diligence!

In the field of interior design, as for a stage that highlights the practical value, only institutions of higher education remain obligated to research and think. Whether it is basic research or in-depth research on a practical case, it is done in the disciplinary field. This is not only precious, but indispensable. As a public organization devoted to academic research, we should do such work, so we must offer support and help for their long-term efforts.

Memorial to the Throne for an Expedition contains many insightful overviews of matter handling, such as "…consult him about all matters, big and small, before acting, and this can necessarily make good defects and gain more benefits." In the field of interior design, the temporal relationship between research and practice is always on the spiral rise. Practice is the start of research, while research is a preparation for massive practice. I always think that interior design has a bright future, and will become an existent form of the architectural design, which has lasted for thousands of years. We need to "seek advice from others and accept admonitions" to make more analyses and studies of interior design, as well as more discussions and prospects on its future.

Su Dan
Written in Milan on 5/9/2018

前言

陈静勇
执行主编 | Executive Editor
中国建筑学会室内设计分会副理事长
Vice President of China Architecture Society
Interior Design Branch
北京建筑大学 教授
Professor of Beijing University of Civil Engineering
and Architecture

不负新时代 担起新使命 迈向新征程
Living up to the New Era and Marching Towards a New Journey with a New Mission on Shoulders

今年是早立秋，节气虽然过了，但如每年一样，载着"室内设计6+"2018（第六届）联合毕业设计丰硕成果的《国粹赓续卷——历史建筑保护与再利用设计》的编制工作仍正在暑热天气中紧张地进行，寓意蒸蒸日上吧！这是"室内设计6+"联合毕业设计特色教育创新项目连续专辑的第6卷，也是中国建筑学会室内设计分会（IID-ASC）第八届理事会教育工作的新开篇。

在建设创新型国家和人才强国战略的指引下，融入我国"一流大学和一流学科建设"的新环境，面对设计人才培养的新目标，服务行业发展的新需求，由中国建筑学会室内设计分会主办，以同济大学、哈尔滨工业大学、华南理工大学、西安建筑科技大学、北京建筑大学、南京艺术学院、浙江工业大学为核心组合的，地处不同区域，设置建筑学或设计学等学科室内设计与理论方向的高校，联合知名设计企业，共同创立的"室内设计6+"联合毕业设计特色教育创新项目，在协同探索室内设计师卓越培养之路上已走过了6年，成为设计创新教育项目品牌平台，彰显出不断丰富的新内涵和不断提升的影响力。

2017年9月20日，住房城乡建设部下发《关于加强历史建筑保护与利用工作的通知》（建规〔2017〕212号），指出充分认识保护历史建筑的重要意义，提出了"做好历史建筑的确定、挂牌和建档；最大限度发挥历史建筑使用价值；不拆除和破坏历史建筑；不在历史建筑集中成片地区建高层建筑"等加强历史建筑的保护与利用指导意见。

来自建筑学、环境设计、工业设计、艺术与科技

Autumn began earlier this year. Despite this, the compilation of *National Quintessence Continuation—Historic Building Conservation and Reuse Design*, which carries the fruitful achievements made in the "interior Design 6+" 2018 (Sixth Year) Joint Graduation Project Event, remains in full swing the intense heat of summer as before. The hot weather may mean increasing prosperity! This is the 6th volume of the Innovation Program of Characteristic Education on "interior Design 6+" Joint Graduation Project, and a new chapter in the educational work undertaken by the eighth council of the Institute of Interior Design-ASC (IID-ASC).

Guided by the strategy of making China an innovative nation and revitalizing China through talents development, in the new context of "construction of world-class universities and subjects", faced with the new objective of designer training and the new requirements of the growing service industry, the Innovation Program of Characteristic Education on "interior Design 6+" Joint Graduation Project was founded 6 years ago by the IID-ASC with universities that are located in different places but offer architecture or design-related interior design and theory, including Tongji University, Harbin Institute of Technology, South China University of Technology, Xi'an University of Architecture and Technology, Beijing University of Civil Engineering and Architecture, Nanjing University of the Arts and Zhejiang University of Technology, in collaboration with well-known design companies in order to jointly explore a way of cultivating eminent designers, becoming a branded platform of design innovation education, showing a constantly enriched new connotation and continuously improved influence.

On September 20th, 2017, the Ministry of Housing and Urban-Rural Development issued *Notice on strengthening Historic Building Conservation and Reuse (No.212 [2012])*, pointing out the great significance of fully understanding historic building conservation and offering guidance on strengthening historic building conservation and reuse by suggesting "identifying, licensing and archiving historic buildings", giving full play to the use value of historic buildings, not demolishing or destroying historic buildings, and not building high-rise buildings where there are many historic buildings".

A comprehensive practical teaching activity is carried out under the theme of " National Quintessence Continuation—Historic Building Conservation and Reuse Design" for the graduates of 2018 majoring in architecture, environment design, industrial design, art and technology through collaboration among the IID-ASC, universities, enterprises and experts, to make an interdisciplinary exploration on historic buildings and their values, the idea of historic building

等专业的2018届毕业生们，通过室内分会、高校、企业、专家之间的协同，就"国粹赓续——历史建筑保护与再利用设计"总命题下的子课题方向，联合开展了综合性实践教学活动，对多维视野下的历史建筑及其价值、历史建筑保护和再利用的历史与理念、历史建筑再利用及其新旧关联模式、新旧空间关联及其相关建筑理论等多个领域的共同关注点等方面的探讨感同身受，并进行深入交流，促进了毕业设计教学水平和人才培养质量的提升。其中，同济大学的建筑学专业师生，以"同济大学图书馆室内外环境保护与再生设计"为题，开展高校图书馆历史建筑保护与再利用设计探讨。

哈尔滨工业大学的建筑学、环境设计两个专业的师生，分别以"文化寄居——哈尔滨道里文化宫改造设计""历史建筑保护背景下建筑空间及其社区环境改造设计"为题，开展哈尔滨道里历史街区文化宫历史建筑保护与再利用设计探讨。

华南理工大学的环境设计专业师生，分别以"华工附属中学旧建筑再生——高校众创空间设计""30个模块+7个技术——云县传统民居更新设计"为题，开展附中旧建筑再利用设计和传统民居更新设计探讨。

西安建筑科技大学的环境设计专业师生，以"韩城古城的保护与再利用设计""山西省夏门村梁氏古堡建筑群修复与再利用设计"为题，开展历史城市、古村落建筑保护与更新设计探讨。

北京建筑大学的工业设计专业师生，分别以"恭王府博物馆展览设计·广式家具制作技艺精品展""恭王府博物馆（忻州工作站）展览设计·晋作家具制作技艺精品展"为题，开展北京、山西两地文物建筑保护与展陈利用设计探讨。

南京艺术学院的艺术与科技专业师生，以"万花筒——民国时期人物展"为题，开展南京民国总统府（太平天国天王府）文物建筑保护与展陈利用设计探讨。

浙江工业大学建筑学、环境设计两个专业的师生，分别以"有机更新，新旧共生""合·游"为题，开展浙江省薛下庄村"二十四间"老建筑保护与再利用设计探讨。

本届联合毕业设计项目主要环节的"6+"是指：

（1）命题研讨。2017年11月9日（南昌市，南昌大学）中国建筑学会室内设计分会2017第二十七届（江西）年会："室内设计6+"2018（第六届）联合毕业设计命题研讨会。

conservation and reuse, historic building reuse and its new-old association pattern, new-old spatial correlation and its architectural theories from multiple perspectives. The vicarious exploration and in-depth exchange of views can promote the improvement of the graduation project teaching level and talent training quality. The teachers and students in architecture from Tongji University made discussion on the conservation and reuse of historic university library buildings under the title of "Indoor and Outdoor Environmental Conservation and Renovation Design of Tongji University Library".

The teachers and students in architecture and environment design from Harbin Institute of Technology made a discussion on the conservation and reuse design of the historic cultural palace building in Daoli District, Harbin City under the title of "Cultural Sojourn—Renovation Design of Harbin Daoli Cultural Palace" and "Renovation Design of an Architectural Space and its Community Environment in the Context of Historic Building Conservation".

The teachers and students in environment design from South China University of Technology made a discussion on the reuse design of the old buildings in the attached middle school and the renovation design of traditional dwellings under the title of "Regeneration of the Old Buildings in the Attached Middle School of South China University of Technology—Design of a Public Entrepreneurial Space at University" and "30 Modules+7Technologies—Renovation Design of Yun County Traditional Dwellings".

The teachers and students in environment design from Xi'an University of Architecture and Technology made a discussion on the conservation and renovation design of historic cities and ancient village buildings under the title of "Conservation and Reuse Design of the Historic Buildings in the Ancient City of Hancheng" and "Renovation and Reuse Design of Liang Family Castle Building Complex in Xiamen Village, Shanxi Province".

The teachers and students in industrial design from Beijing University of Civil Engineering and Architecture made a discussion on the conservation and display design of the historic buildings in Beijing and Shanxi under the title of "Prince Kung's Palace Museum—Exhibition Design for Guangdong-style Furniture Craftsmanship and Masterpieces" and "Prince Kung's Palace Museum (Xinzhou Work Station)—Exhibition Design for Guangdong-style Furniture Craftsmanship and Masterpieces".

The teachers and students in art and technology from Nanjing University of the Arts made a discussion on the conservation and display design of the historic buildings in the Nanjing Presidential Palace of the Republic of China (the Heavenly Palace of the Taiping Heavenly Kingdom) under the title of "Kaleidoscope—Personages of the Republican Period".

The teachers and students in architecture and environment design from Zhejiang University of Technology made a discussion on the conservation and ruse design of the "twenty-four" old buildings in Xuexiazhuang Village, Zhejiang Province under the title of "Organic

（2）开题报告。2018年3月9日—11日（哈尔滨市，哈尔滨工业大学）进行开题报告、专家讲坛和历史建筑考察。

（3）中期检查。4月27日—29日（南京市，南京艺术学院）进行中期检查、专家讲坛和历史建筑考察。

（4）答辩评审。6月27日—29日（上海市，同济大学）进行答辩评审、专家讲坛、总结与表彰奖励。

（5）编辑出版。中国建筑学会室内设计分会和北京建筑大学负责汇集参加高校、支持企业、专家讲坛与点评寄语等，主编中国建筑学会室内设计分会推荐专业教学参考书："室内设计6+" 2018（第六届）联合毕业设计《国粹赓续卷——历史建筑保护与再利用设计》，并由中国水利水电出版社于（重庆）年会期间出版发行。

（6）专题展览。10月27日—29日"室内设计6+" 2018（第六届）联合毕业设计专题展览在中国建筑学会室内设计分会2018年二十八届（重庆）年会暨国际学术交流会期间举办。

"+"指对外交流。这是本项目的1个特色拓展环节。《国粹赓续卷——历史建筑保护与再利用设计》主要内容采用中英文对照方式编辑出版，是作为中国建筑学会室内设计分会（IID-ASC）与亚洲室内设计联合会（AIDIA）等开展设计教育国际交流的特色案例之一。

《国粹赓续卷——历史建筑保护与再利用设计》编制工作继续探索基于本项目前5届专辑初步形成的书籍设计特色。本卷"编制栏目"的6+涵盖了相应项目环节的核心内容。其中，"调研踏勘""中期检查""答辩展示""教育研究""专家讲坛""风采定格"6个栏目记录了本届项目的主体内容和过程，附有专家点评、学生感言、获奖证书、师生照片等；第1栏目的"项目规章"中，汇编有项目"章程""答辩、评审、表彰工作细则""纲要""框架任务书""书籍设计"等，是项目有章可循，开展长效机制建设的体现。

响应国家创新驱动发展战略和服务经济社会发展需求，培养造就一大批创新能力强、适应需要的高质量室内设计师，始终是全社会等共同面对的命题。

"室内设计6+" 2018（第六届）联合毕业设计特色教育创新项目已顺利完成。从联合毕业活动中走来的建筑学、工业设计、环境设计、艺术与科技等专业的毕业生们，获得了相应的建筑学学士（专业学位）、工学学士、艺术学学士学位，他们继而有的考取了硕士研究生，

Renewal and Co-existence of the Old and the New" and "Heyou".

"6+", the main part of the present joint graduation project event refers to:

(1) Discussion on topic assignment. The 27th (Jiangxi) Annual Meeting of the IID-ASC (at Nanchang University, Nanchang City) on November 9th 2017: The Topic Assignment Seminar on the "interior Design 6+" 2018 (Sixth Year) Joint Graduation Project Event.

(2) Opening report. Opening report, expert forum and historic building research (at Harbin Institute of Technology, Harbin City) on March 9th—11th, 2018.

(3) In-process inspection. In-process inspection, expert forum and historic building research (at Nanjing University of the Arts, Nanjing City) on April 27th—29th.

(4) Oral defense review. Oral defense review, expert forum, summarization, commendation and rewards (at Tongji University, Shanghai City) on June 27th—29th.

(5) Compilation and publication. The IID-ASC and Beijing University of Civil Engineering and Architecture are responsible for compiling the professional teaching reference book: "interior Design 6+" Joint Graduation Project Event *National Quintessence Continuation—Historic Building Conservation and Reuse Design* with materials collected from the universities, enterprises and experts involved. Then the book will be published by China Water&Power Press and distributed during the annual meeting (in Chongqing).

(6) Special exhibition. On October 27th—29th, the special exhibition on the "interior Design 6+" 2018 (Sixth Year) Graduation Project Event was held during the 28th (Chongqing) Annual Meeting of IID-ASC 2018 and International Academic Exchange Conference.

"+"— External exchange. It is a featured extension of the program. The main content of *National Quintessence Continuation* is written in Chinese and English. It is one of the characteristic cases of the international exchange on design education between the IID-ASC and AIDIA.

The National Quintessence Continuation is compiled based on the book design features formed in the past 5 years. The 6+ "compiling columns" consist of the core content of the relevant "event parts". Of the parts, "on-the-spot survey" "in-process inspection" "oral defense display" "education research" "expert forum" and "mien freeze-framing" record the main body and process content of this year's event, attached with expert comments, students' words, honor certificates and teachers' and students' photos; in "Event Regulations" in the first part there are the "Event Charter" "Outline" "Framework Assignment" and "Detailed Regulations on Oral Defense, Review and Commendation". This shows that the event has rules to follow and can carry out long-term mechanism construction.

For the whole society, it is always necessary to train a great many high-quality interior designers with strong innovation ability to meet the requirements of the national innovation-driven development strategy and economic and social development.

有的出国留学深造，有的已就业开始工作等，共同踏上了面向行业需求、走向卓越发展的新征途。

今年是中国建筑学会成立65周年，"室内设计6+"联合毕业设计特色教育创新项目走过了6年的探索历程。在此，以《国粹赓续卷——历史建筑保护与再利用设计》作为纪念版呈献给朋友们。

希望"室内设计6+"联合毕业设计特色教育创新项目为"一流大学和一流学科建设"发挥出促进作用。根据相关代表性特色高校条件及其组合意愿情况，逐步在我国六大区（华北、华东、中南、东北、西南、西北地区）增设"中国（×地区）'室内设计6+'联合毕业设计"，在各省市增设"中国（×省／市）'室内设计6+'联合毕业设计"。进一步拓展品牌交流平台，提升我国设计人才联合培养质量。

感谢全国高等学校建筑学学科专业指导委员会、教育部高等学校设计学类专业教学委员会长期以来对高校相关学科专业建设工作的指导！

感谢相关高校、命题与支持企业、讲座与评审专家、出版企业等对"室内设计6+"联合毕业设计项目的支持和帮助！

感谢该项目参加高校的鼎力承办，以及志愿者们的辛勤付出！

2017年11月，中国建筑学会室内设计分会完成了换届工作，产生了第八届理事会。"室内设计6+"联合毕业设计特色教育创新项目也将承前启后、继往开来，不负新时代，担起新使命，迈向新征程。

由于时间限制，编制中难免挂一漏万，敬请指正，以利改进。

陈静勇
二〇一八年八月七日 （戊戌·立秋）

The Innovation Program of Characteristic Education on "interior Design 6+" Joint Graduation Project has come to a successful end. The students in architecture, industrial design, environment design and art and technology involved in the event have graduated with a bachelor's degree of architecture, engineering or arts. Some of them have now been postgraduate students, some have gone abroad for further studies and some have started their career. All of them have set foot on a new journey to excellent development to meet the industry's needs.

This year marks the 65th anniversary of the Architectural Society of China, and the Innovation Program of Characteristic Education on "interior Design 6+" Joint Graduation Project has been founded for 6 years. Here, I would like to take *National Quintessence Continuation* as a souvenir to all my friends.

I hope that the Innovation Program of Characteristic Education on "interior Design 6+" Joint Graduation Project will facilitate the "construction of world-class universities and subjects". The "China (×region) 'Interior Design 6+' Joint Graduation Project Event" and "China (×province/city) 'Interior Design 6+' Joint Graduation Project Event" can be carried out gradually in the six regions of China (North China, East China, South Central China, Northeast China, Southwest China and Northwest China) and all provinces and cities respectively in accordance with related representative characteristic universities' conditions and will to combine with one another, to further expand the branded exchange platform to improve the quality of the joint training of designers.

Thank the Professional Guiding Committee for Architecture at Chinese Universities, the Ministry of Education and other professional design teaching guidance committees concerned for their long-term guidance on the construction of the related disciplines and professions at the universities!

Thank the universities involved, topic assigners, corporate supporters, lecturers, reviewers and publisher for their support and help for the "interior Design 6+" Joint Graduation Project Event!

Thank the universities for their generous participation and the volunteers for their hard work!

In November 2017, the IID-ASC changed the term of office and formed the eighth council. The Innovation Program of Characteristic Education on "interior Design 6+" Joint Graduation Project will take over from the past and set a new course for the future, carry forward the cause pioneered by our predecessors and forge ahead into the future to live up to the New Era and march towards a new journey with a new mission on shoulders.

Due to time limitations, there may be omissions, so please correct us, so that we could improve the content.

Chen Jingyong
August 7th, 2018 (The First Day of Autumn)

恭王府博物馆（忻州工作站）展览设计·晋作家具制作技艺精品展

万花筒——民国时期人物展 140

合·游——浙江省薛下庄村『二十四间』老建筑保护与再利用设计 146

同济大学图书馆室内外环境保护与再生设计 154

30个模块+7个技术——云县传统民居更新设计 162

历史建筑保护背景下建筑空间及其社区环境改造设计 170

韩城古城的保护与再利用设计 178

恭王府博物馆展览设计·广式家具制作技艺精品展 186

有机更新，新旧共生——浙江省薛下庄村『二十四间』老建筑保护与再利用设计研究 192

五 教育研究 208

中国建筑学会室内设计分会『"室内设计6+"联合毕业设计特色教育创新项目』报告（2013届—2017届） 200

2018同济大学图书馆保护与更新改造毕设指导心得 左琰 林怡 210

建筑学专业毕业设计成果评价模式的探讨 周立军 马辉 兆翚 216

环境设计专业本科毕业设计联合指导模式探索 刘晓军 王敏 220

恭王府博物馆文化空间探索 杨琳 陈静勇 李奕慧 223

多元联合教学模式的探讨——由『室内设计6+』联合毕业设计教学活动引发的思考 王一涵 227

六 专家讲坛 236

宋微建 韩冠恒 朱海玄

七 风采定格 242

目 录

序　出师表　006

前言　不负新时代，担起新使命，迈向新征程　008

一　项目规章　014

『室内设计6+1』联合毕业设计章程（2018修订版）　016

『室内设计6+1』2018（第六届）联合毕业设计框架任务书　021

『室内设计6+1』2018（第六届）联合毕业设计项目纲要　025

『室内设计6+1』联合毕业设计答辩、评奖、表彰工作细则（2018修订版）　028

『规矩』说——《国粹赓续卷》书籍设计　032

二　调研踏勘　034

三　中期检查　074

四　答辩展示　102

同济大学图书馆室内外环境保护与再生设计　104

华工附属中学旧建筑再生——高校众创空间设计　114

文化寄居——哈尔滨道里文化宫空间改造设计　124

山西省夏门村梁氏古堡建筑群修复与再利用设计　132

热点命题,纷显特色

联合指导,服务需求

项目规章

《室内设计6+》2018（第六届）
联合毕业设计
Interior Design 6+ 2018 (Sixth Year)
University-Enterprise Cooperative Graduation Project Event

课题方向
历史街区与建筑、城市场所再生
陶瓷艺术酒店、文化旅游项目设计
辽宁营口河北地、便落古镇设计

"室内设计 6+"联合毕业设计章程（2018 修订版）

"Interior Design 6+" Joint Graduation Project Event Charter (2018)

　　为服务城乡建设领域室内设计专门人才培养需求，加强室内设计师培养的针对性，促进相关高等学校在专业教育教学方面的交流，引导面向建筑行（企）业需求开展综合性实践教学工作，由中国建筑学会室内设计分会（以下简称"室内分会"）倡导并主管，国内外设置室内设计相关专业（方向）的高校与行业代表性建筑与室内设计企业开展联合毕业设计。

　　为使联合毕业设计活动规范有序，形成活动品牌和特色，室内分会在征求相关高等学校意见和建议的基础上形成最初的《中国建筑学会室内设计分会（CIID）"室内设计 6+1"联合毕业设计章程》，并于 2013 年 1 月 13 日室内分会（CIID）"室内设计 6+1" 2013（首届）联合毕业设计（北京）命题会上审议通过，公布试行，后逐年修订。

　　历经 2013—2017 年连续五届联合毕业设计的深入交流，原 CIID "室内设计 6+1"联合毕业设计取得了丰富成果，形成一定影响力，积累了室内分会设计教育平台建设成功经验。2017 年 10 月，室内分会第八届理事会通过《教育工作规划纲要（2017 年—2025 年）》，将该活动提升为今后持续开展的"室内设计 6+"联合毕业设计特色教育创新项目，更名为"室内设计 6+"联合毕业设计，编制《"室内设计 6+"联合毕业设计章程》，并公布试行。

一、联合毕业设计设立的背景、目的和意义

　　2010 年教育部启动了"卓越工程师教育培养计划"，于 2011—2013 年分三批公布了进入"卓越计划"的本科专业和研究生层次学科。2011 年国务院学位委员会、教育部公布《学位授予和人才培养学科目录（2011 年）》，增设了"艺术学（13）"学科门类，将"设计学（1305）"设置为"艺术学"学科门类中的一级学科。"环境设计"建议作为"设计学"一级学科下的二级学科，"室内设计"建议作为新调整的"建筑学（0813）"一级学科下的二级学科。2012 年教育部公布《普通高等学校本科专业目录（2012 年）》，在"艺术学"学科门类下设"设计学类（1305）"专业，"环境设计（130503）"等成为其下核心专业。"艺术学"门类的独立设置，设计学一级学科以及环境设计、室内设计等学科专业的设置与调整，形成了我国环境设计教育和室内设计专门人才培养学科专业的新格局。

　　2015 年 10 月，国务院发布《统筹推进

　　To meet the demand for professional interior design talent training in the urban and rural construction field, strengthen the pertinence of interior designer cultivation, promote related universities to exchange ideas with one another on professional education and teaching, and assist the universities in carrying out comprehensive practice teaching to meet the needs of the construction industry (enterprises), the domestic and foreign universities that offer interior design-related specialties and representative architectural and interior design enterprises cooperate in doing graduation project design on the initiative of the China Institute of Interior Design (IID-ASC).

　　In order that the joint graduation project event should go with a swing in an orderly manner, the CIID formulated the former CIID "Interior Design 6+1" Joint Graduation Project Event Charter based on opinions and suggestions from the universities involved. Later, the charter was deliberated and approved at the topic-assignment meeting of the CIID "Interior Design 6+1" 2013 (First Year) Joint Graduation Project Event in Beijing on January 13th, 2013. Then, the charter was put into operation on trial and is amended annually.

　　After in-depth exchange of views on the joint graduation project event in the past 5 years from 2013 to 2017, the former "Interior Design 6+1" Joint Graduation Project Event achieved fruitful results, accumulating successful experience in the construction of an educational platform for interior design. In October 2017, the 8th council of the CIID passed the Educational Planning Framework (2017—2025), upgraded the event to Innovation Project of Characteristic Education on "interior Design 6+" Joint Graduation Project, renaming it "interior Design 6+" Joint Graduation Project Event, and compiled "interior Design 6+" Joint Graduation Project Event Charter, which was then put into operation on trial.

I. The Background, Objective and Significance of the Joint Graduation Project Event

　　In 2010, the Ministry of Education initiated the "Excellent Engineer Training Program"; in 2011-2013, it made public the undergraduate specialties and postgraduate programs listed in the "Excellent Program" in three times. In 2011, the Academic Degree Commission of the State Council and the Ministry of Education released the Catalogue of the Degree and Talent Training Subjects (2011), additionally offering the "art (13)" specialty, setting the "design (1305)" as a first-level topic in the "art" specialty. The "environment design" was proposed as a second-level topic subordinate to the "design", and the "Interior design" was proposed as a second-level topic subordinate to the "architecture (0813)", which had just been upgraded to be a second-level subject. In 2012, the Ministry of Education released the Catalogue of the Undergraduate Programs offered at Regular Institutions of Higher Education (2012), putting the "design (1305)" and "environment design" under the category of the "art" as core specialties. The setup and adjustment of the "art", as well as the design, a first-level subject, and the interior design formed a new pattern of professional interior design talent training in the area of environment design education.

世界一流大学和一流学科建设总体方案》。2017年1月，教育部、财政部、国家发展改革委印发《统筹推进世界一流大学和一流学科建设实施办法（暂行）》。党的十九大报告指出："建设教育强国是中华民族伟大复兴的基础工程，必须把教育事业放在优先位置，加快教育现代化，办好人民满意的教育。""一流大学和一流学科建设"是建设高等教育强国、实现十九大提出的"实现社会主义现代化和中华民族伟大复兴"总任务的必然选择和重要举措。

因此，组织开展室内设计领域联合毕业设计，对加强相关学科专业特色建设，深化综合性实践各教学环节交流，促进室内设计教育教学协同创新，培养服务行（企）业需求的室内设计专门人才，具有十分重要的意义。

二、联合毕业设计组织机构

1. 指导单位和主办单位

"室内设计6+"联合毕业设计由室内分会主办，受全国高等学校建筑学学科专业指导委员会、教育部高等学校设计学类专业教学指导委员会等指导。

2. 参加高校、承办高校和总（参）编高校

联合毕业设计一般由学科专业条件相近，设置室内设计方向相关专业的6所高校间通过协商、组织成为活动参加高校。应突出参加高校组合的地理区域、办学类型、专业特色、就业面向等方面的代表性和涵盖性，在学科专业间形成一定的既具有交叉性又具有协同设计的工作环境和交流氛围。

发挥"室内设计6+"联合毕业设计特色教育创新项目在"双一流建设"中促进作用，可以根据相关代表性特色高校条件及其组合意愿情况，逐步在我国六大区（华北、华东、中南、东北、西南、西北地区）增设"（×地区）'室内设计6+'联合毕业设计"，在各省市增设"（×省/市）'室内设计6+'联合毕业设计"。

每年通过参加高校申报和室内分会遴选，确定毕业设计开题报告、中期检查、答辩评审等3场集中活动的承办高校，以及《（主题）×卷——（总命题）×》[中国建筑学会室内设计分会推荐专业教学参考书："室内设计6+"×（年）（第×届）联合毕业设计]（以下简称《主题卷》）总编高校，其他参加高校作为参编高校。

每所高校参加联合毕业设计到场汇报

In October 2015, the State Council issued An Overall Plan on Comprehensively promoting the Construction of World-class Universities and Subjects. In January 2017, the Ministry of Education, the Ministry of Finance and the National Development and Reform Commission printed and issued Measures for Comprehensively promoting the Construction of World-class Universities and Subjects (Interim). The report of the 19th CPC National Congress indicates: "It is a foundation project related to the rejuvenation of China to build China into a great power of education. This necessitates making the educational cause a priority, quickening educational modernization and meeting the people's requirements for education." The "construction of world-class universities and subjects" is an inevitable choice and important measure to build China into a great power of higher education and fulfill the overall task of "realizing socialist modernization and the rejuvenation of China" set at the 19th CPC National Congress.

Therefore, the joint graduation project of interior design is of great significance to strengthening the characteristic construction of relevant subjects and specialties, deepening exchanges on comprehensive practical teaching, promoting collaborative innovation of interior design education and teaching, and training professional interior design talents required for service industries (enterprises).

II. The Organizer of the Joint Graduation Project Event

1. Guider and organizer

The "interior Design 6+" Joint Graduation Project Event is organized by the IID-ASC, and guided by the Professional Guiding Committee for Architecture at Chinese Universities, the Ministry of Education and other various professional design teaching guidance committees.

2. Universities involved, Co-organizer, and Chief Compiler (Co-compilers)

The joint graduation project event is generally attended by 6 universities that offer similar subjects and specialties, including the interior design, through consultation. These universities' geographic region, teaching type, professional characteristics and employment orientation should be highlighted, to hold different disciplines together, and create an environment and atmosphere of communication for collaborative design.

To give full play to the role of the Characteristic Educational Innovation Program on "interior Design 6+" Joint Graduation Project in "double first-class construction", the (×region) "Interior Design 6+" Joint Graduation Project Event and (×province/city) "Interior Design 6+" Joint Graduation Project Event can be carried gradually out in the six regions of China (North China, East China, South Central China, Northeast China, Southwest China and Northwest China) and all provinces and cities respectively in accordance with related representative characteristic universities' conditions and will to combine with one another.

The annual co-organizer of opening report, in-process inspection and defense review, as well as the chief compiler of the (Subject)×[volume]—(General Assignment)× [the professional teaching reference book recommended by the IID-ASC: "interior Design 6+"×(Year) (the xth) Joint Graduation Project] (hereinafter referred to as "the Topic Volume"), is

学生一般以6人为宜，分为2个设计方案组；要求配备1～2名指导教师，其中至少有1名指导教师具有高级职称；高校导师熟悉环境设计、室内设计等工程实践业务，与相关领域企业联系较广泛。室内分会负责聘任参加高校导师，参与联合毕业设计工作。

3. 命题企业

参加高校向室内分会推荐所在省（市、地区）的行业代表性建筑与室内设计企业作为毕业设计命题企业，企业命题人应具有高级职称；室内分会负责聘任企业命题人作为联合毕业设计企业导师。企业导师与相应高校导师协同编制联合毕业设计总命题下的《"（分课题）×"毕业设计教学任务书》，参与联合毕业设计相关环节工作。

4. 支持企业

参加高校向室内分会推荐行业代表性建筑与室内设计企业作为毕业设计支持企业；由室内分会与支持企业签订活动支持与回馈协议，负责聘任支持企业观察员，参与联合毕业设计相关环节工作。

5. 出版企业

室内分会和《主题卷》总编高校遴选业知名出版企业，作为联合毕业设计《主题卷》出版企业，并参与联合毕业设计相关环节工作。

三、联合毕业设计流程环节

（1）联合毕业设计每年举办1届，与参加高校毕业设计教学工作实际相结合。

（2）室内分会负责联合毕业设计总体策划、宣传，组织研讨、编制、公布每届联合毕业设计《〈主题〉×——（总命题）×框架任务书》、《"室内设计6+"2018（第六届）联合毕业设计项目纲要》，制订、公布《"室内设计6+"联合毕业设计答辩、评价工作细则》等，协调参加高校、命题企业、相关机构等，聘请课题专家开展专题学术讲座，负责对毕业设计课题成果质量、毕业设计优秀指导教师、毕业设计优秀组织单位、毕业设计特殊贡献等内容进行评价，以及组织室内设计教育国际交流等活动。

（3）联合毕业设计主要教学环节包括命题研讨、开题报告、中期检查、答辩评审及课题评价、《主题卷》编辑出版、专题展览等6主要环节，对外交流作为联合毕业设计活动的1个扩展环节。相关工作分别由室内分会、参加高校、命题企业、支持企业、出版企业等分工协同落实。

（4）命题研讨。室内分会组织召开联

selected from the universities involved by the IID-ASC, while the other universities serve as co-compilers.

Each university shall have 6 students attend the joint graduation project event and be present to make a report, and they shall be subdivided into 2 design teams; 1~2 supervisors, including at least one with a senior professional title, take part in the event with the students; university supervisors are familiar with engineering practice, including environment design and interior design, and stay in touch with many enterprises concerned. The IID-ASC is obliged to engage supervisors from the universities involved and have them participate in the joint graduation project event.

3. Topic Assigner

Each university involved shall recommend to the IID-ASC a representative provincial (municipal or regional) interior design enterprise as the topic assigner, and the enterprise shall have a senior professional title; the IID-ASC shall hire the topic assigner as a corporate supervisor of the joint graduation project event. The corporate supervisor shall assist the university supervisor(s) concerned in compiling a "(sub-topic)×" graduation project teaching assignment under the General Assignment, and guiding the joint graduation project.

4. Corporate Supporter

Each university involved shall recommend to the IID-ASC a representative architectural and interior design enterprise as a supporter for the joint graduation project event; the IID-ASC shall sign a event support and feedback agreement with the supporter and hire it as an inspector to participate in the joint graduation project event.

5. Publishing Enterprise

The IID-ASC and the chief compiler of the Topic Volume shall select a well-known publishing enterprise as the Topic Volume and a participant in the joint graduation project event.

III. Joint Graduation Project Process

1. The join graduation project event, held once a year, is in line with the universities' graduation project teaching.

2. The IID-ASC shall make overall planning for the joint graduation project event, publicize it, organize discussions, compile files, release the (Subject)×—(General Assignment)×Framework Assignment and Event Outline, make and release Guidelines on "interior Design 6+" Joint Graduation Project Defense and Evaluation", and so on. Also, it shall assist the universities, topic assigners and other agencies in hiring professional experts to hold an academic forum, and organize a evaluation on the quality of the design results, excellent graduation project supervisors, excellent graduation project participants and special contributions to graduation project design, as well as international communication on interior design education.

3. The process of joint graduation project teaching consists of 6 parts: topic assignment discussion, in-process inspection, defense review and topic evaluation, compilation and publication of the Topic Volume, and special exhibition, while external communication is an extension of the joint graduation project event. The above work shall be done by the IID-

合毕业设计命题研讨会。每届联合毕业设计的总命题着眼室内设计等相关领域学术前沿和行业发展热点问题，参加高校协同命题企业细化总命题方向的下分课题。联合毕业设计分课题要求具备相关设计资料收集、现场踏勘、建设管理方支持等条件。

命题研讨会一般安排在高校秋季学期中（每年11月左右），结合当年室内分会年会安排专题研讨。

（5）开题报告。室内分会组织开展联合毕业设计开题活动，颁发联合毕业设计高校导师和企业导师聘书，承办高校协同安排开题活动启动、专家学术讲座、开题汇报与专家点评、调研参观等。每所参加高校合组报告开题情况，其中分命题介绍与开题报告陈述不超过20分钟，专家点评不超过10分钟。

开题活动一般安排在高校春季学期开学初（3月上旬）进行。

（6）中期检查。室内分会组织开展联合毕业设计中期检查活动，承办高校协同安排专家学术讲座、中期成果汇报与专家点评、调研参观等活动。每所参加高校优选不超过2个过程方案组进行中期检查，其中每组陈述不超过20分钟，专家点评不超过10分钟。

中期检查一般安排在春季学期期中（4月下旬）进行。

（7）答辩评审与课题评价。室内分会组织开展联合毕业设计答辩评审及课题评价活动，承办高校协同安排答辩评审、课题评价等工作。每所参加高校优选不超过2个答辩方案组进行陈述与答辩、成果展出，其中每组陈述不超过20分钟，专家点评与学生回答不超过10分钟。

在答辩、成果展示、评审的基础上，室内分会组织开展对《室内设计6+》联合毕业设计特色教育创新项目》的年度课题评价，重点评价毕业设计课题成果质量、毕业设计优秀指导教师、毕业设计优秀组织单位、毕业设计特殊贡献等。坚持"质量第一、宁缺毋滥"的评审原则，毕业设计课题成果质量评价结果分为优秀、良好、合格、不合格四个等级，其中优秀、良好评价一般按照1：2比例设置。毕业设计课题成果质量评价仅针对答辩方案设置，评价结果等级可以空缺。

答辩评审及课题评价一般安排在春季学期期末（6月上旬）进行。

（8）专题展览。室内分会在每届联合

ASC, universities, topic assigners, supporters and publishing enterprises separately or synergistically.

4. Topic Assignment Discussion

The IID-ASC shall organize a discussion on topic assignment for the joint graduation project event. The general assignment of each year's joint graduation project event shall be focused on the academic frontier and industrial hot spots of interior design, and the universities involved shall make a subtopic under the general assignment. The subtopic of the joint graduation project event requires design data acquisition, reconnaissance trip, and support from construction management.

Generally, the topic assignment discussion is held around the annual topic in the middle autumn term (about November every year).

5. Opening Report

The IID-ASC shall organize an opening report meeting on the joint graduation project event, award a letter of appointment to university supervisors and corporate supervisors assist in holding a academic forum, making opening speeches and comments, and organize research and visits. Every university shall give one opening report, in which subtopic introduction and opening report presentation shall take not more than 20min, and expert comments shall take not more than 10min.

Generally, an opening report is given in the beginning of the spring term (early March).

6. In-process Inspection

The IID-ASC shall organize an in-process inspection on joint graduation project design, to assist university experts in holding an academic forum, report in-process achievements, and organize research and visits. Every university involved shall choose up to 2 process scheme teams for in-process inspection, and either team shall give a presentation that takes not more than 20min, and each expert shall spend not more than 10min to make comments.

Generally, in-process inspection is conducted in the middle spring term (late April).

7. Defense Review and Topic Evaluation

The IID-ASC shall organize a defense review and topic evaluation activity for the joint graduation project event in collaboration with the universities involved. Every university involved shall choose up to 2 oral defense teams for presentation, oral defense and a display of achievements. To be specific, either team shall give a presentation of up to 20min; expert comments and student answers shall take up to 10min.

On the basis of defense, achievement display and review, the IID-ASC shall organize an annual evaluation on the topic of the Innovation Project of Characteristic Education on "interior Design 6+" Joint Graduation Project, to primarily evaluate the quality of the design results, as well as excellent graduation project supervisors, excellent graduation project participants and special contributions to graduation project design. Review shall follow the principle of "quality first and quality superior to quantity". The results of the evaluation on the quality of the topic achievements are divided into four levels: excellent, good, qualified and unqualified, of which the ratio of excellent to good is 1:2. The quality evaluation is for

毕业设计结束当年的室内分会年会暨学术研讨会（每年10—11月）举办期间安排联合毕业设计作品专题展览；专题展览结束后，相关高校可自愿向室内分会申请联合毕业设计作品巡回展出。

（9）编辑出版。基于每届联合毕业设计成果，由室内分会组织编辑出版《主题卷》，作为室内分会推荐的专业教学参考书。《主题卷》由室内分会和总编高校、参编高校联合编著，参加高校导师负责排版和审稿等工作，出版企业作为责任编辑，负责校审、出版、发行等工作。

（10）对外交流。室内分会和出版企业一般在每届联合毕业设计结束当年室内分会年会期间联合举行《主题卷》发行式；由室内分会联系亚洲室内设计联合会（AIDIA）等，开展室内设计教育成果国际交流，宣传中国室内设计教育，拓展国际交流途径。

四、联合毕业设计相关经费

（1）室内分会负责筹措对毕业设计课题成果质量、毕业设计优秀指导教师、毕业设计优秀组织单位、毕业设计特殊贡献等的评价经费，以及室内分会年会专题展览、宣传等环节经费。

（2）参加高校自筹参加联合毕业设计各环节经费。

（3）承办高校负责联合毕业设计开题报告、中期检查、答辩评审与课题评价等环节的宣传海报、场地、设备等工作；答辩评审环节承办高校还负责答辩作品展出场地、展板制作等经费；《主题卷》总编负责出版等主要经费。

（4）命题企业、支持企业、出版企业等负责为向校企联合毕业设计提供一定形式的支持等。

五、附则

本章程于2018年3月12日公布试行，由中国建筑学会室内设计分会负责解释。原《中国建筑学会室内设计分会（IID-ASC）"室内设计6+1"联合毕业设计章程》同时废止。

oral defense only, and the real evaluation results may not necessarily at all the four levels.

Generally, defense review and topic evaluation are performed in the late spring term (early June).

8. Special Exhibition

The IID-ASC organizes a special exhibition of joint graduation project contents after the joint graduation project event ends and during the period of the annual meeting and academic conference; after the end of the special exhibition, the universities concerned can voluntarily apply to the IID-ASC for an itinerant exhibition of joint graduation project contents.

9. Compilation and Publication

The IID-ASC organizes the compilation and publication of the Topic Volume based on each year's joint graduation project results as a professional teaching reference book. The Topic Volume shall be co-compiled by the IID-ASC, the chief compiler and the other universities involved, and each university's supervisors shall be responsible for examining its own manuscript, while the publishing enterprise shall serve as an editor in charge for proofreading and publication.

10. External Exchanges

Generally, the IID-ASC and publishing enterprise co-distribute the Topic Volume after the end of the joint graduation project event and during the IID-ASC's annual meeting; the IID-ASC shall communicate with the Asia Interior Design Institute Association (AIDIA) on the results of interior design education, propagandize Chinese interior design education and expand the way of international communication.

IV. Costs of Joint Graduation Project

1. The IID-ASC is responsible for raises funds for evaluation on the quality of the design results, as well as excellent graduation project supervisors, excellent graduation project participants and special contributions to the graduation project, as well as for the special exhibition and propaganda.

2. Costs of universities' participation in the joint graduation project event.

3. The universities shall prepare posters, spaces and equipment for the opening report, in-process inspection, defense review and topic evaluation; also, the universities shall fund the display of exhibition and the production of display panels; the chief compiler of the Topic Volume shall fund the publication.

4. The topic assigner, supporter and publishing enterprise shall offer some support to the joint graduation project event.

V. Supplementary Provisions

The charter shall be put into operation on trial on March 12th, 2018, and be interpreted by the IID-ASC. The former IID-ASC "Interior Design 6+1" Joint Graduation Project Event Charter shall be abolished simultaneously.

"室内设计 6+" 2018（第六届）联合毕业设计框架任务书

"interior Design 6+" 2018 (Sixth Year) Joint Graduation Project Event Framework Assignment

《"室内设计 6+" 2018（第六届）联合毕业设计框架任务书》（简称《2018 框架任务书》）是依据中国建筑学会室内设计分会（以下简称室内分会）2017 第二十七届（江西）年会命题研讨会意见，经"室内设计 6+" 2018（第六届）联合毕业设计导师组编制形成。活动参加高校应结合本校毕业设计教学工作实际，协同相应支持企业，据此进一步编制本校相应《"室内设计 6+" 2018（第六届）联合毕业设计详细任务书》（简称《2018 详细任务书》），指导毕业设计教学工作，开展联合毕业设计活动。

一、总命题

历史建筑是指经城市、县人民政府确定公布的具有一定保护价值，能够反映历史风貌和地方特色的建筑物、构筑物，是城市发展演变历程中留存下来的重要历史载体。近年来，对历史建筑保护与再利用问题研究与实践已成为我国快速城市化发展进程中的焦点和热点之一。加强历史建筑的保护和合理利用，有利于展示城市历史风貌，留住城市的建筑风格和文化特色，是践行新发展理念、树立文化自信的一项重要工作。

为此，中国建筑学会室内设计分会将"国粹赓续——历史建筑保护与再利用设计"作为"室内设计 6+" 2018（第六届）联合毕业设计的总命题。参加高校子课题名称在本校《2018 详细任务书》中拟定。

二、总体原则

1. 历史建筑保护原则

1964 年在联合国教科文组织倡导下提出的《威尼斯宪章》，推进了全世界的历史建筑保护工作。1982 年 11 月 19 日第五届全国人民代表大会常务委员会第二十五次会议通过《中华人民共和国文物保护法》；2017 年 11 月 4 日第十二届全国人民代表大会常务委员会第三十次会议第五次修正。

（1）必须原址保护。
（2）尽可能减少干预。
（3）定期适时日常保养。
（4）保存现存事物原状与历史信息。
（5）按照保护要求使用保护技术。
（6）正确把握审美标准。
（7）必须保护文物环境。
（8）已不存在的建筑不应重建。
（9）考古发掘应注意保护实物遗存。
（10）预防灾害侵害。

2. 历史建筑再利用原则

必须坚持以社会效益为准则，不应为了

"interior Design 6+" 2018 (Sixth Year) Joint Graduation Project Event Framework Assignment (hereinafter referred to as "the Framework Assignment 2018") is compiled by the supervisor group of the "interior Design 6+" 2018 (Sixth Year) Joint Graduation Project Event according to what was discussed at the 27th Annual Meeting of the China Institute of Interior Design (IID-ASC) held in Jiangxi. Every university involved should compile a more detailed assignment for its graduation project (hereinafter referred to as "the Detailed Assignment 2018") in cooperation with its corporate supporter based on the Framework Assignment 2018 in accordance with its actual situation to guide graduation project teaching and carry out joint graduation project events.

I. Overall Topic

By a historic building is meant a building or structure that has a certain conservation value and can reflect the historical features and local characteristics of a place as is declared by the municipal or county government. In recent years, China has been focused on the research and practice of historic building conservation and reuse in the process of rapid urbanization. It helps to display a city's historical features and retain its architectural style and cultural characteristics to strengthen the conservation and rational use of historic buildings. It is of great significance to practicing new ideas of development and build cultural self-confidence.

To this end, the China Institute of Interior Design (IID-ASC) designates "National Quintessence Continuation—Historic Building Conservation and Reuse Design" as the overall topic of the "interior Design 6+" 2018 (Sixth Year) Joint Graduation Project Event. Each university involved shall select a subtopic in its Detailed Assignment 2018.

II. General Guidelines

1. Principles for Historic Building Conservation

The Venice Charter, compiled and released on the initiative of the UNESCO in 1964, has been boosting historic building conservation worldwide. The Law of the People's Republic on the Protection of Cultural Relics was passed at the 25th Session of the 5th Standing Committee of the National People's Congress on November 19th, 1982; the Fifth Amendment was passed at the 13th Session of the 12th Standing Committee of the National People's Congress on November 4th, 2017.

(1) Cultural relics shall be conserved where it is.
(2) Intervention shall be minimized.
(3) Operational maintenance shall be implemented regularly in due time.
(4) The original state and historical information of extant cultural relics shall be well preserved.
(5) Conservation technology shall be adopted as required.
(6) Aesthetic standards shall be adopted properly.
(7) The environment of cultural relics shall be protected.
(8) Any building that has disappeared shall not be rebuilt.
(9) Historical remains shall be protected during archaeological excavation.

当前利用的需要而损害文物古迹的价值。确定为文物保护单位的历史建筑，根据《中华人民共和国文物保护法》按原状保存，不能损毁、改建、添建或者拆除，维修和保养要体现"整旧如故"的原则。使用上一般可作为博物馆、保管所或参观场所。

三、项目地点

参加高校遵循所在省市区域中历史文化街区保护规划，在文物类、保护类、改善类建筑名录中考察、遴选代表性历史建筑的项目地点。

四、设计范围

依据历史建筑保护规划划定的历史建筑保护范围、建设控制地带范围、缓冲区范围。

五、设计内容

毕业设计方案应基于《2018框架任务书》总命题和总体原则，体现保护性利用设计专业内容，将对历史建筑的保护与再利用统一起来。各校在《2018详细任务书》中明确具体设计内容和要求。

六、设计深度

（1）开题调研汇报：历史建筑调研报告、设计概念分析、《国粹赓续卷》开题调研内容排版页。

（2）中期检查汇报：初步设计方案、《国粹赓续卷》中期检查内容排版页。

（3）毕业答辩汇报：深化设计方案、答辩展板、《国粹赓续卷》毕业答辩内容排版页。

七、设计成果

（一）设计说明

设计说明主要包含：历史建筑保护与再利用设计理念、定位、方案设计分析、经济技术指标等图示及图表。

（二）图纸（对应设计内容）

（1）历史建筑区域位置图。

（2）历史建筑保护规划图和规划分析图。

（3）历史建筑利用设计平面图、顶面图、剖（立）面图、代表性详图、分析图。

（4）历史建筑利用设计装置单体平面（俯）、仰）视图、正（侧）视图、剖面图、代表性详图、分析图。

（三）彩色效果图

（1）历史建筑与环境鸟瞰图。

（2）历史建筑利用设计效果图。

（3）历史建筑利用设计装置效果图。

（四）成果提交

(10) Disasters shall be prevented.

2. Principles for Reuse of Historic Buildings

Social benefits shall come first, and the value of cultural relics and historic sites shall not be undermined for the sake of current benefits. The historic buildings identified as officially protected sites shall be preserved in status quo according to the *Law of the People's Republic on the Protection of Cultural Relics*, and shall not be damaged, rebuilt, extended or demolished, and any historic building shall be maintained with its "appearance unchanged". It can be used as a museum, depository or scenic spot.

III. Project Location

Each university involved shall follow the provincial or municipal planning for historic and cultural district conservation and choose a representative historic building from the list of the historic buildings that need to be renovated.

IV. Design Scope

The design scope covers the historic buildings, development control areas and buffer zones designated in the historic building conservation planning.

V. Design Content

The graduation project should be done based on the general assignment and general guidelines in the *Framework Assignment 2018*, to compile content around protective use, to assure the conservation and use of historic buildings. Each university shall create specific design content and requirements in the *Detailed Assignment 2018*.

Ⅵ. Design Depth

(1) Opening report: historic building research report, design concept analysis, and typesetting of the opening report content of *National Quintessence Continuation*.

(2) In-process inspection report: preliminary design scheme, and typesetting of the in-process inspection content of *National Quintessence Continuation*.

(3) Oral defense report: detailed design scheme, defense panel, and typesetting of the oral defense content of *National Quintessence Continuation*.

Ⅶ. Design Results

(I) Design Description

Content of design description: idea and orientation of historic building conservation and reuse design, conceptual design analysis, and design of charts such as economic and technical norms.

(II) Drawing (corresponding to the design content)

(1) Regional location map of the historic building.

(2) Planning chart and planning analysis chart for historic building conservation.

(3) Plane graph, top surface map, profile, representative detailed drawing and analysis chart of historic building reuse.

(4) (Vertical) plane view, front (side) view, sectional view, representative detailed chart and analysis chart of single furnishings in

1. 开题汇报

每所参加高校汇总 1 个汇报文件进行陈述；每校答辩时间限 20 分钟，专家点评 10 分钟。提交开题调研成果 PPT、《国粹赓续卷》开题调研内容排版页等。

2. 中期检查

每所参加高校优选 2 个方案组进行陈述；每组答辩时间限 20 分钟，专家点评 10 分钟。提交中期检查成果 PPT、《国粹赓续卷》中期检查内容排版页等。

3. 毕业答辩

（1）每所参加高校优选 2 个答辩方案组进行陈述；每组答辩时间限 20 分钟，专家点评 10 分钟；提交毕业答辩成果、展板、《国粹赓续卷》毕业答辩内容排版页等。

（2）每个方案设计组的展板限 3 张，展板规格为幅面 A0 加长：900mm×1800mm 竖版，分辨率不低于 100dpi。展板模板由室内分会按照年会展板要求统一提供。展板由毕业设计答辩活动承办高校（同济大学）负责喷绘、布置。

4. 出版素材

为编辑出版《国粹赓续卷——历史建筑保护与再利用设计》[中国建筑学会室内设计分会推荐专业教学参考书："室内设计 6+"2018（第六届）联合毕业设计]，活动相关参加单位和个人等需积极响应室内分会相关工作要求，具体内容如下：

（1）单位简介（参加高校、指导企业、支持企业各 1 篇，中文 1000 字以内，中英文对照；单位标识矢量文件）。

（2）教研论文（每所高校导师联名 1 篇，中文 3000 字以内，中英文对照）。

（3）开题调研排版（每所高校"开题调研"内容排版占 2 页或 4 页，主要标题和关键词等需要中英文对照）。

（4）中期检查排版（每所高校 2 个方案"中期检查"内容排版各占 2 页或 4 页，主要标题和关键词等需要中英文对照）。

（5）答辩作品排版（每所高校每个方案在"设计内容"的两个方面排版共占 6 页或 8 页，主要标题和关键词等需要中英文对照）。

（6）讲座提要（每位讲座专家 1 篇，

reused historic buildings.

(III) Color Picture

(1) Bird's-eye view of the historic building and surroundings.

(2) Design sketch of historic building reuse.

(3) Design sketch of reused historic building.

(IV) Achievements Submission

1. Opening Report

Every university involved shall submit a summary report of presentation; every university will be given 20min for oral defense, and experts will be given 10min to make comments. Submit opening research achievements in PPT, and typesetting of the opening research content of *National Quintessence Continuation*.

2. In-process Inspection

Every university will choose 2 scheme groups for presentation; each group will spend 20min on oral defense, and the experts will spend 10min to make comments. Submit in-process inspection achievements in PPT, and typesetting of the in-process inspection content of *National Quintessence Continuation*.

3. Oral Defense

(1) Every university shall form 2 oral defense teams, either of which shall spend 20min on oral defense, and the experts shall spend 10min to make comments; submit oral defense results, a display panel, and typesetting of the oral defense content of *National Quintessence Continuation*.

(2) Each team can have 3 panels at most, and each panel shall be lengthened based on breadth A0: 900mm×1800mm vertical, resolution ratio≥100dpi. The panel template will be offered by the IID-ASC as requested. The organizer (Tongji University) will be responsible for spray painting and decoration.

4. Material Publication

To edit and publish *National Quintessence Continuation—Historic Building Conservation and Reuse Design* (the IID-ASC recommends the following professional teaching reference book: "interior Design 6+" 2018 (Sixth Year) Joint Graduation Project Event), the participants, including universities and individuals, shall actively meet the requirements of the IID-ASC:

(1) Self-introduction (about each university itself, as well as its corporate guider and supporter, not more than 1000 Chinese characters, in both Chinese and English; vector files of logo).

(2) teaching research thesis (each university shall submit 1 thesis jointly written by the supervisors, not more than 3000 Chinese characters, in both Chinese and English).

(3) Typesetting of opening research (each university shall use 2P or 4P to contain opening research, with the main titles and keywords in both Chinese and English).

(4) Typesetting of in-process inspection (each university shall use 2P or 4P to contain in-process inspection, with the main titles and keywords in both Chinese and English).

(5) Typesetting of oral defense (each university shall use 6P or 8P to contain the "designed content" of either scheme, with the main titles and keywords in both Chinese and English).

(6) Lecture synopsis (each lecturer shall make a speech, not more than

中文 1500 字以内，中英文对照）。

（7）专家寄语（每位专家 1 段，中文 300 字以内，中英文对照）。

（8）专家点评（专家、导师点评相应方案 1 段，中文 300 字以内，中英文对照）。

（9）学生感言（每个方案组 1 段，中文 200 字以内，中英文对照）。

（10）工作照片（每位专家、导师、学生各 1 张）。

（11）评审证书（室内分会提供，证书电子版）。

（12）活动照片（各校归集提供，各主要环节照片电子版）。

（13）答辩 PPT（每所高校 2 个方案答辩 PPT 文件）。

（14）作品展板（每所高校 2 个方案作品展板 TIF 原文件）。

八、附建筑与场地图

具体内容见各校《2018 详细任务书》。

1500 Chinese characters, in both Chinese and English).

(7) Words from experts (each expert shall make a speech, not more than 300 Chinese characters, in both Chinese and English).

(8) Experts' comments (each expert or supervisor shall make a comment, not more than 300 Chinese characters, in both Chinese and English).

(9) Words from students (each team shall make a speech, not more than 200 Chinese characters, in both Chinese and English).

(10) Work photos (each expert, supervisor and student shall provide 1 photo).

(11) Review certificate (the IID-ASC shall provide a review certificate, as well as an electronic one).

(12) Event photos (the universities involved shall provide electronic event photos).

(13) PPT files of oral defense (each university shall provide 1 PPT file of the 2 oral defenses).

(14) Work display panel (each university shall provide the original file in TIF for the display of the 2 schemes).

Ⅷ. Building Drawing and Floor Plan

See each university's *Detailed Assignment 2018*.

"室内设计6+" 2018（第六届）联合毕业设计项目纲要

依据《"室内设计6+"联合毕业设计章程》和《"室内设计6+" 2018（第六届）联合毕业设计命题研讨会纪要》，经与参加高校协商，形成本届联合毕业设计项目纲要，指导活动开展。

一、总命题： 国粹赓续——历史建筑保护与再利用设计

二、项目地点： 参加高校遵循所在省市区域中历史文化街区保护规划，在文物类、保护类、改善类建筑名录中考察、遴选代表性历史建筑的项目地点。

三、指导单位： 全国高等学校建筑学学科专业指导委员会
　　　　　　　教育部高等学校设计学类专业教学指导委员会

四、主办单位： 中国建筑学会室内设计分会

五、承办高校： 南昌大学［中国建筑学会室内设计分会2017第二十七届（江西）年会："室内设计6+" 2018（第六届）联合毕业设计命题研讨会］
　　　　　　　哈尔滨工业大学（开题踏勘、专家讲坛、历史建筑考察）
　　　　　　　南京艺术学院（中期检查、专家讲坛、历史建筑考察）
　　　　　　　同济大学（答辩评审、专家讲坛、展览、总结与表彰奖励）

六、主编单位： 中国建筑学会室内设计分会
　　　　　　　北京建筑大学

七、参加高校组合（学院\专业）：
　　　　　　　同济大学（建筑与城市规划学院\建筑学）
　　　　　　　华南理工大学（设计学院\环境设计）
　　　　　　　哈尔滨工业大学（建筑学院\建筑学、环境设计）
　　　　　　　西安建筑科技大学（艺术学院\环境设计）
　　　　　　　北京建筑大学（建筑与城市规划学院\工业设计）
　　　　　　　南京艺术学院（工业设计学院\艺术与科技）
　　　　　　　浙江工业大学（建筑工程学院\建筑学、设计艺术学院\环境设计）

八、指导企业： 参加高校依据总命题遴选确定历史建筑保护与再利用设计项目依托单位，作为本校联合毕业设计指导企业（一校一企）

九、特邀观察员： 亚洲城市与建筑联盟（AAUA）、亚洲设计学年奖组委会

十、出版企业： 中国水利水电出版社

十一、媒体支持： 中国室内设计网 http://www.ciid.com.cn

十二、时　　间：2017年11月—2018年10月

十三、活动安排：

序号	阶段	时间（年/月/日）	地点	活动内容	相关工作
1	命题研讨	2017/11/09	分会二十七届（江西）年会南昌大学（研讨会承办）	● 11月8日，报到，年会场外活动 ● 11月9日—10日，分会二十七届（江西）年会 ● 11月9日，2018（第六届）联合毕业设计命题研讨会	● 总结交流往届活动经验 ● 研讨"室内设计6+" 2018（第六届）联合毕业设计总命题和各校子课题 ● 确定本届活动承办高校

序号	阶段	时间（年/月/日）	地点	活动内容	相关工作
2	教学准备	2017/11/11—2018/03/08	各高校	● 联合毕业设计教学工作准备	● 各高校报送参加毕业设计师生名单 ● 各高校反馈对《活动纲要》的修改意见和建议 ● 各高校邀请联合企业（一校一企业），依据《框架任务书》，分别编制和向学会报送本校《毕业设计详细任务书》 ● 各高校安排文献检索与民居考察毕业实习 ● 编制开题调研汇报PPT、书稿开题调研分配页面电子版 ● 准备开题仪式和民居考察
3	开题踏勘	2018/03/09—2018/03/11	哈尔滨工业大学	● 3月9日，报到 ● 3月10日，开题仪式、专题讲座、调研交流 ● 3月11日，历史建筑保护与再利用考察	● 开题仪式；分会颁发导师聘书 ● 专题讲座；分会颁发讲座专家聘书 ● 开题调研汇报；提交开题调研成果 ● 商讨中期检查、《国粹赓续卷》编辑出版等相关工作 ● 历史建筑保护与再利用考察
4	方案设计	2018/03/12—2018/04/18	参加高校	● 方案设计	● 参加高校安排相关讲授、考察、辅导、设计、研讨等活动 ● 明确方案设计目标，完成相关文案、图表等编制 ● 完成调研报告、方案设计基本图示、效果图、模型、分析图表及等中期成果 ● 编制中期检查方案设计汇报PPT、《国粹赓续卷》中期检查页面 ● 准备中期检查和历史建筑保护与再利用考察
5	中期检查	2018/04/27—2018/04/29	南京艺术学院	● 4月27日，报到 ● 4月28日，中期检查 ● 4月29日，历史建筑保护与再利用考察	● 中期检查汇报；提交中期检查成果 ● 专家点评，形成方案设计深化重点 ● 商讨答辩评审、表彰奖励、编辑出版、展览等相关工作 ● 历史建筑保护与再利用考察
6	深化设计	2018/04/23—2018/06/14	参加高校	● 方案深化设计	● 完成方案深化设计图示、效果图、模型、分析图表、设计说明等相应成果；编制展板 ● 编制毕业设计答辩PPT、书稿答辩方案分配页面排版 ● 准备答辩评审、历史建筑保护与再利用考察、表彰奖励
7	答辩评审	2018/06/22—2018/06/24	同济大学	● 6月22日，报到 ● 6月23日，答辩、评审 ● 6月24日，历史建筑保护与再利用考察；表彰奖励	● 答辩布展与观摩；提交《国粹赓续卷》答辩方案成果 ● 毕业答辩；等级奖评审 ● 表彰联合毕业设计项目方案、联合毕业设计优秀导师、联合毕业设计最佳组织单位、联合毕业设计突出贡献单位等 ● 2018（第六届）活动总结

序号	阶段	时间（年/月/日）	地点	活动内容	相关工作
8	编辑出版	2018/06/18—2018/10/20	北京市	● 7月15日前，高校完成《国粹赓续卷》作品部分排版工作，提交分会秘书处 ● 9月5日前，完成书稿总编工作，提交出版企业校审 ● 9月20日前，完成书稿校审工作，送印厂 ● 10月20日前，《国粹赓续卷》印刷完成	● 参加高校和相关专家提交《国粹赓续卷》其他页面内容 ● 分会和北京建筑大学负责总编《国粹赓续卷》 ● 出版企业负责书稿校审 ● 分会2018第二十八届（重庆）年会"室内设计6+"2018（第六届）联合毕业设计命题研讨会准备
9	展览交流	2018/10/27—2018/10/29	重庆市	● 10月27日，报到 ● 10月27日，"室内设计6+"2018（第六届）联合毕业设计命题研讨会暨《国粹赓续卷》出版发行式 ● 10月27日—29日，年会专题展览	● 展板布展与观摩 ● 发行《国粹赓续卷》 ● 研讨确定"室内设计6+"2019（第七届）联合毕业设计命题 ● "室内设计6+"2018（第六届）联合毕业设计专题展览 ● 亚洲室内设计联合会（AIDIA）设计教育成果交流

"室内设计 6+"联合毕业设计答辩、评奖、表彰工作细则（2018 修订版）

中国建筑学会室内设计分会（以下简称室内分会）依据《"室内设计 6+"联合毕业设计章程》（2018 版），制订"室内设计 6+"联合毕业设计答辩、评审、表彰工作细则，指导相关单位和人员遵照执行。

一、答辩准备

（1）参加高校每校到场指导教师不超过 2 名，到场学生不超过 6 名。

（2）参加高校每校安排不超过 2 个毕业设计答辩方案组，参加毕业设计答辩评审和成果展出；此外，每校最多可再报送 2 个不参加毕业设计答辩评审的自荐方案，仅参加毕业设计成果展出。

（3）每个毕业设计答辩方案组提前准备毕业设计答辩陈述 PPT 等电子文档，于毕业设计答辩活动报到时提交活动组委会。

（4）答辩前每个毕业设计答辩方案、自荐方案需分别准备成果展板 3 张；使用室内分会统一发布的模板编辑，展板幅面为 A0 加长：900mm×1800mm，分辨率不低于 100dpi。展板电子版须于答辩前 1 周发送到答辩活动承办高校指定的工作邮箱；由承办高校负责汇总打印、布展等。

（5）参加高校按当届《"室内设计 6+"联合毕业设计书稿排版要求》编辑书稿，于毕业设计答辩活动现场提交室内分会活动组委会。

（6）毕业设计课题方向选报。"室内设计 6+"联合毕业设计参加高校，应结合本校毕业设计教学实际，按照当届《"室内设计 6+"联合毕业设计"×（主题）——×（总命题）"框架任务书》（以下简称《框架任务书》）设置的毕业课题方向，选报每个毕业设计答辩方案组的课题方向（单选），并于答辩前 1 周报送室内分会活动组委会。

二、答辩与评审

（1）每届毕业设计答辩评委会由特邀评委和高校评委组成。

1）特邀评委一般由室内学会专家、命题企业、活动观察员、支持企业等在内的 5～7 位专家担任；答辩评委会组长一般由学会遴选的专家（总导师）担任。

2）高校评委由参加高校各推选 1 位毕业设计指导教师担任。

（2）毕业设计答辩评审。

1）毕业设计答辩评审结果按当届毕业设计课题方向分别设置优秀、良好、合格、不合格 4 个等级，优秀和良好的比例一般为 1∶2，评审等级结果可以空缺。

Detailed Regulations on Oral Defense, Review and Commendation for "interior Design 6+" Joint Graduation Project Event (2018)

The China Institute of Interior Design (IID-ASC) makes the Detailed Regulations on Oral Defense, Review and Commendation for the "interior Design 6+" Joint Graduation Project Event in accordance with the "interior Design 6+" Joint Graduation Project Event Charter for the participants to follow and carry out.

I. Preparation for Oral Defense

1. For each university involved in the joint graduation project event, it shall send a maximum of 2 supervisors and 6 students to the event.

2. Each university shall form up to 2 defense scheme teams for a review of oral defense and a display of achievements. In addition, each university shall submit a maximum of 2 self-recommended schemes to be displayed rather than orally defended.

3. Every defense scheme team shall prepare an electronic statement of defense in PPT or another format and submit it to the organizing committee while signing up for the oral defense activity.

4. Before the oral defense begins, 3 achievement display panels shall be prepared for each defense scheme or self-recommended scheme; the scheme content shall be typeset according to the edit model released by the IID-ASC, and the panel size is larger than A0, namely 900mm×1800mm, with a resolution of no less than 100dpi. Please send an electronic panel to the e-mail box specified by the organizer 1 week in advance of the oral defense; the organizer shall be responsible for printing all the electronic panels and arranging exhibitions.

5. The universities involved shall compile a book in accordance with the Requirements for "interior 6+" Joint Graduation Project Design Manuscript Typesetting and then submit it to the IID-ASC before the oral defense begins.

6. Graduation Project Topic Assignment. Every university involved in the "interior 6+" Joint Graduation Project Event shall assign a graduation project topic for either defense scheme team under the present "interior 6+" Joint Graduation Project Event "× (Topic)—× (General Topic)" (hereinafter referred to as Framework Assignment) based on its practical graduation project teaching and submit it to the organizing committee 1 week in advance of oral defense.

II. Oral Defense and Review

1. The oral defense jury is composed of guest reviewers and reviewers from the universities

(1) Generally, 5~7 guest reviewers are invited, and they may be experts from the IID-ASC, cooperative or supporting enterprises, or event inspectors; the jury leader is usually an expert (chief superior) appointed by the IID-ASC.

(2) For the appointment of the reviewers from the universities involved, 1 supervisor at each university shall be selected.

2. Oral Defense Review

(1) According to the topic assignment, the review results of oral defense are divided into four levels: excellent, good, qualified and unqualified, of which the ratio of excellent to good is 1:2. The real review results may not necessarily be at all the four levels.

(2) The first round of review. Each university's either defense scheme

2）第一轮评选。参加高校每个毕业设计答辩方案组按选报课题方向进行答辩，每组答辩陈述时间不超过15分钟，问答不超过15分钟。由特邀评委、高校评委共同填写选票，进行排序评选（如，1为建议排序第一，2为建议排序第二等，依次类推）。活动组委会负责排序选票统计，形成相应设计专题"毕业设计等级奖"评选建议排序。

3）第二轮评选。高校评委须回避，由特邀评委以"毕业设计等级奖"评选建议排序为基础，对照答辩方案进行审议评选，确定当届"毕业设计等级奖"获奖方案。

（3）"毕业设计佳作奖"评选。

答辩评委会对参加成果展示的高校自荐方案展板进行评议，对是否认定为"毕业设计佳作奖"方案进行投票；同意票数超过答辩评委会总人数1/2（含）的高校自荐方案，确定获得佳作奖。

（4）"优秀毕业设计导师"评选。

获得"毕业设计等级奖""毕业设计佳作奖"作品的命题企业总导师、参加高校导师成为当届"优秀毕业设计导师"。

（5）"毕业设计最佳组织单位"评选。

承办当届联合毕业设计活动开题踏勘、中期检查、毕业答辩、总辑出版的高校成为"毕业设计最佳组织单位"。

（6）"毕业设计突出贡献单位"评选。

负责当届联合毕业设计活动的命题企业、重要支持企业等成为"毕业设计突出贡献单位"。

三、表彰奖励

（1）在活动颁奖典礼上，由室内分会领导分别向"毕业设计等级奖""毕业设计佳作奖""优秀毕业设计导师""毕业设计最佳组织单位""毕业设计突出贡献单位"的获得者颁发证书。

（2）奖励证书由学会盖章有效；"毕业设计等级奖""优秀毕业佳作奖"证书印有活动答辩评委会评委签名，以示纪念。

四、附则

（1）本规则2013年6月8日公布施行，由中国建筑学会室内设计分会负责解释。

（2）本规则2015年5月第一次修订，2016年5月第二次修订，2017年6月第三次修订，2018年6月第四次修订。

group makes an oral defense for its topic, and every team's defense time will not exceed 15min. The guest reviewers and reviewers from the universities shall cast votes together and make a review in sequence (e.g., 1 represents No.1, 2 represents No.2, and the like). The organizing committee is responsible for counting votes to create a propositional sequence to decide on "graduation project level awards".

(3) The second round of review. Only the guest reviewers are qualified to make a review of the defense schemes according to the propositional sequence to decide on "graduation project level awards".

3. "Excellent Works"

The oral defense jury shall make a review of the universities' self-recommended scheme panels and then cast a vote to determine whether the panels should be named "excellent works"; for a self-recommended scheme panel, it shall be named an excellent work if half or more than half of the reviewers make an affirmative vote for it.

4. "Excellent Graduation Project Supervisors"

The chief supervisors from the corporate topic assigners and universities whose works are award the "graduation project grade prize" and named "excellent works" shall be named "excellent graduation project supervisors".

5. "The best organizer of the graduation project event"

"The best organizer of the graduation project event" shall be the university responsible for compiling an opening report, making an in-process inspection, organizing an oral defense and publishing the final results" at the joint graduation project event.

6. "Outstanding Contributors to the Graduation Project Event"

The candidates are among the corporate topic assigners and supporting enterprises at the joint graduation project event.

III. Commendations and Awards

(1) At the awards ceremony, the leaders of the IID-ASC will present honor certificates to "graduation project grade prize" winners, "excellent works" "excellent graduation project supervisors" "the best organizer of the graduation project event" and "the outstanding contributors to the graduation project event".

(2) The honor certificates will be effective on signing by the IID-ASC; the "graduation project grade prize" and the award for "excellent works" are printed with the reviewers' signature as souvenirs.

IV. Supplementary Provisions

(1) The present regulations were released and put in force on June 8th, 2013, and interpreted by the IID-ASC.

(2) The present regulations were revised for the first time in May 2015, for the second time in May 2016, for the third time in June 2017 and for the fourth time in June 2018.

1. 书籍的外衣设计

"规矩"说——《国粹赓续卷》书籍设计
书籍封面封底封书脊设计
Design instruction

当书籍与读者初次见面时，封面的设计将直接影响到读者对其的瞬间印象。因此，在着手这本书籍的封面设计之前，先对书籍内容进行充分的了解，在此基础上定位到一个最能引起读者关注且满足读者视觉需要的表现元素——书名字体。

在设计书名字体时，选用美术体形式，其不仅个性，还具备普遍的适用性。美术体又分为规则和不规则两种形态。国粹赓续四个字上半部分选用规则形式，强调了字体的规整性，为阅读带来了极大的辨识性；下半部分选用不规则形式，结合历史建筑剪影强调变化性，突出本书主题，也呈现出更完美的视觉效果。

在封底设计中，由于较多的读者都会有拿起一本书先看封底的习惯，因此在封底中会置入关键的信息，即条形码、国际书号、书的价格、责任编辑以及项目标志，方便读取重要信息。

在书脊设计中，置入的主要内容为书名、册别和出版社，作用是当书籍在立置存放时，便于归类和查找。

在本书籍中存在两种文字的需求，一种是具有提示性的引导文字，它能够激发读者的视觉关注和兴趣；一种是类似于正文形式的说明性文字，它起着阐述和细节描述的作用。因此在本书籍中引导性文字选取拓展字体，说明性文字选取基本字体。

在书籍中，字号大小和色彩构成对于文字版式是极其重要的，其不仅可以调节版面氛围，也方便读者阅读。因此在字号选取上，主要选取9点和11点作为正文两级字号，而引导文字的字号根据版面的大小选取了18点、21点等来调节版面氛围。

封面设计

封底设计

书脊设计

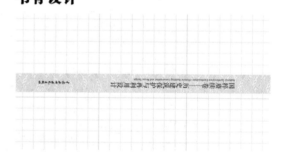

整体版式设计

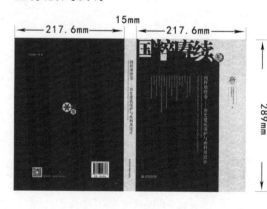

张可凡[1]、王艳[1]、叶红[2]、陈静勇[1T]

1 北京建筑大学设计艺术研究院
2 中国建筑学会室内设计分会
T 通讯作者：北京建筑大学设计艺术研究院

2·书籍的内在美设计

书籍版式设计的构成要素
Design instruction

文字元素

基本字体：
宋体 Regular
ABCDEFGHIJKL
MNOPQRSTUVWX
YZ
abcdefghijkl
mnopqrstuvwx
yz
黑体
ABCDEFGHIJKL
MNOPQRSTUVWX
YZ
abcdefghijkl
mnopqrstuvwx
yz
Calibri Regular

拓展字体：
方正清刻本悦宋简体
迷你简粗宋
文悦古典明朝體

字号标准
国粹赓续　　　　　　　　　　　　　　8
National Quintessence

国粹赓续　　　　　　　　　　　　　　9
National Quintessence

国粹赓续　　　　　　　　　　　　　　18
National Quintessence

国粹赓续　　　　　　　　　　　　　　21
National Quintessence

国粹赓续　　　　　　　　　　　　　　24
National Quintessence

色彩元素

主体色

C5 M6 Y13 K0

C18 M19 Y26 K0

C48 M98 Y97 K23

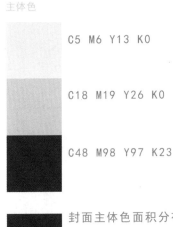

封面主体色面积分布

隔页主体色面积分布

辅助色

C63 M53 Y48 K1

C24 M42 Y77 K0

C82 M44 Y62 K2

C92 M85 Y54 K25

C80 M75 Y72 K48

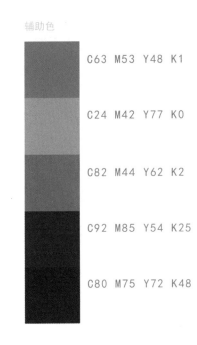

國粹賡續卷——歷史建築保護與再利用設計

項目規章

031

书籍版式设计的构成要素
Design instruction

2. 书籍的内在美设计

在这个读图时代，一本好的图书往往是图文并茂，而图形无疑是抓住读者视线的最有利武器。因此在设计图形时，主要选取正方形、圆形和三角形为基础元素，以减法为处理手法，将三个原始图形延伸为第二层次基础图形，即长方形，圆环和直角三角形；再做减法和扭曲，延伸出第三层次基础图形，即等腰三角形，直线和曲线，三层基础图形元素将作为创意拓展图形的基础元素源头。

在进行一、二、三等数字的图形设计时，首先选取一个正方形为基础元素，以三角、方形和直线三个元素根据文字特征对正方形进行减法切割，得出最终数字拓展图形。其和圆环及肌理相结合，突出方圆规整风格，不仅容易吸引读者的视线，也成为视觉的第一落脚点；其次，拓展图形相比于单纯的文字更具有美感和趣味，让读者不易产生疲倦乏味的感觉。

图形元素

基础图形元素：

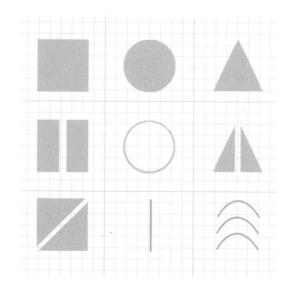

拓展图形：

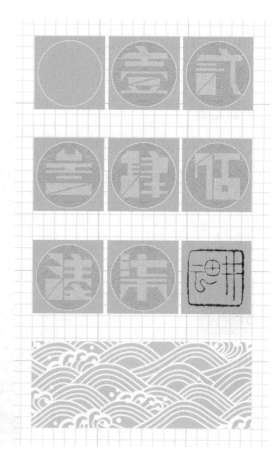

2. 书籍的内在美设计

Design instruction

书籍版式设计的基本框架

隔页版式

正文版式

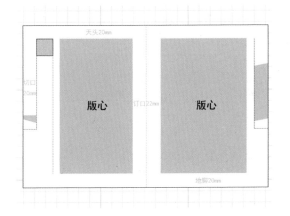

照片版式

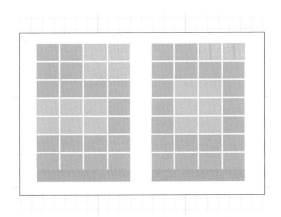

在本书中，由于一些图形较多，因此选用较大版心，甚至采用出血的处理方法，使整个版面更加的生动活跃，增加视觉的流动性，同时避免元素之间产生的拥挤感。在天头和地脚设计中，采取相等的留白处理，不仅显得现代主流，也体现出整本书的规整性。

内文纸张的选择上，主要采用微黄的彩色胶版纸，突出本卷主题的文化和品质，烘托出别样雅致的效果。

文字在隔页中，主要以竖式排版，文字与图形相融合于版面，构成点线面三要素。使整个版式更加干净整洁，而文字不仅仅充当了说明性文字，亦成为图形的一部分。

文字在正文中，除引导性文字采取竖式排版，与封面及隔页版式相一致，其他说明性文字以横式排版，行间距选取12，方便阅读。

國粹賡續卷——歷史建築保護與再利用設計

项目规章

热点命题,纷显特色

联合指导,服务需求

调研踏勘

《室内设计学》2018（第八届）
联合毕业设计
Interior Design & 2018 (Sixth Year)
University Interpreate Cooperative Graduation Project Event

阆中古城
深入实地，精细考察，体验建筑与环境，
收集整理资料，研究本次毕业设计的课题

同济大学
Tongji University

同济大学图书馆室内外环境保护与再生设计
Conservation and Regeneration Design of the Indoor and Outdoor Environment of Tongji University Library

「室内设计 6+」2018（第六届）联合毕业设计
"Interior Design 6+"2018(Sixth Year) Joint Graduation Project Event

小组成员
李一丹　马一茗　潘蕾
张迪凡　刘雨婷　张奕晨
许可　毛燕　DANA

校园基地分析 | Campus Base Analysis

同济大学图书馆位于同济大学正门以西。与正门、毛主席像以及之后的大礼堂形成了非常具有仪式以及纪念性意义的轴线关系。

同时，同济大学图书馆在入口处与南北两侧的南北楼以及国立柱也形成了非常有纪念性价值的入口广场，图书馆也像打开的书本，欢迎着同济大学来来往往的人群。

由于图书馆在同济大学所处的重要地理位置以及历史地位，同济大学图书馆共有三个对外出入口，主要出入口正对学校正门，位于整体轴线位置，两个辅助入口主要服务于来自宿舍楼以及食堂的学生群体，位于西侧玻璃廊的两头。

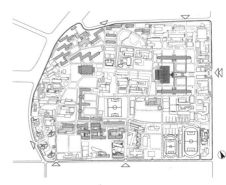
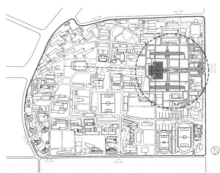

图书馆历史变迁 | The History of the Library

1960 年	1980 年	1990 年	2002 年
			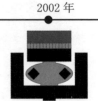
U形阅览楼 藏书楼 目录大厅 二层砖混结构	加建11层塔楼 钢筋混凝土筒核 悬挑结构八角形 目录大厅	加建最西侧书库 与藏书楼封闭链接	保留最西侧书库外壳 改建为阅览室 与藏书库玻璃廊连接 椭圆型玻璃大厅

图书馆历史元素 | The Historical Elements of the Library

20世纪60年代建筑特征

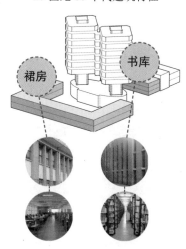

20世纪80年代建筑特征

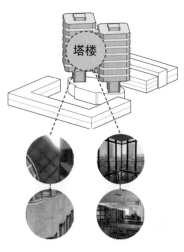

2002年建筑特征

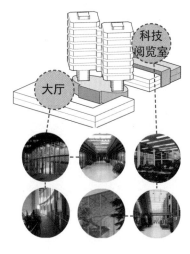

20世纪60年代建造了图书馆U形裙房与书库,两者外立面均质的清水砖墙如今都保留完好,书库内部的钢书架体系也成为了图书馆内重要的历史元素。

20世纪80年代加建了竖直方向的两个核心筒塔楼,其悬挑结构获得了当时的创新奖项,同时外立面的马赛克砖与凸窗也颇具特色。

2002年改造对图书馆植入了玻璃与钢的元素,修建了椭圆形的玻璃大厅,同时对其内部流线也进行了调整。西侧的书库也改为了科技阅览室。

流线现状 | The Status of the Flow Line

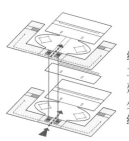

裙房流线过于孤立,与周围建筑联系少,内部动线过长。

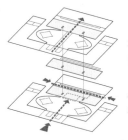

书库处在两条流线的交汇处,内部空间逼仄,流线混杂。

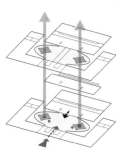

塔楼可达性较差,且两核心筒缺乏标识性,难以分辨方向。

功能现状 | The Status of Functions

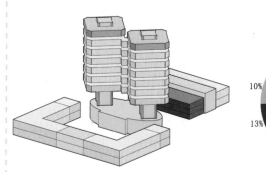

图书馆现状功能分区,大面积都为普通阅览自习教室。

存在问题:
1. 整体功能较为单一,缺乏多样性学生活动场所。
2. 同学们普遍反映很少在图书馆中借书或者阅览书籍,图书馆纸质书本使用率偏低,阅览现状堪忧。

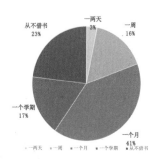

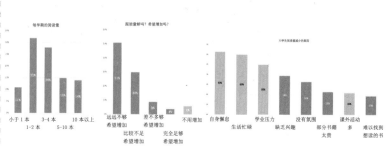

附件：建筑与场地图

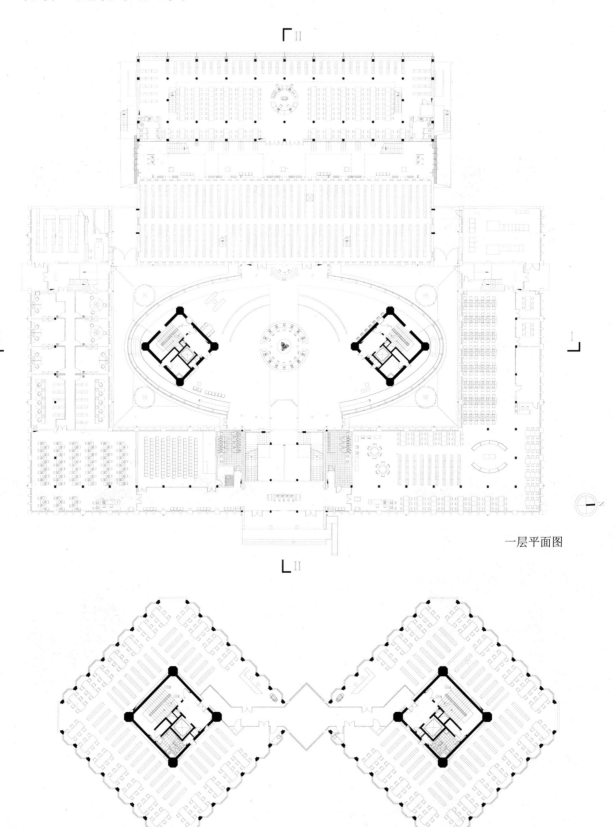

一层平面图

塔楼标准层平面图

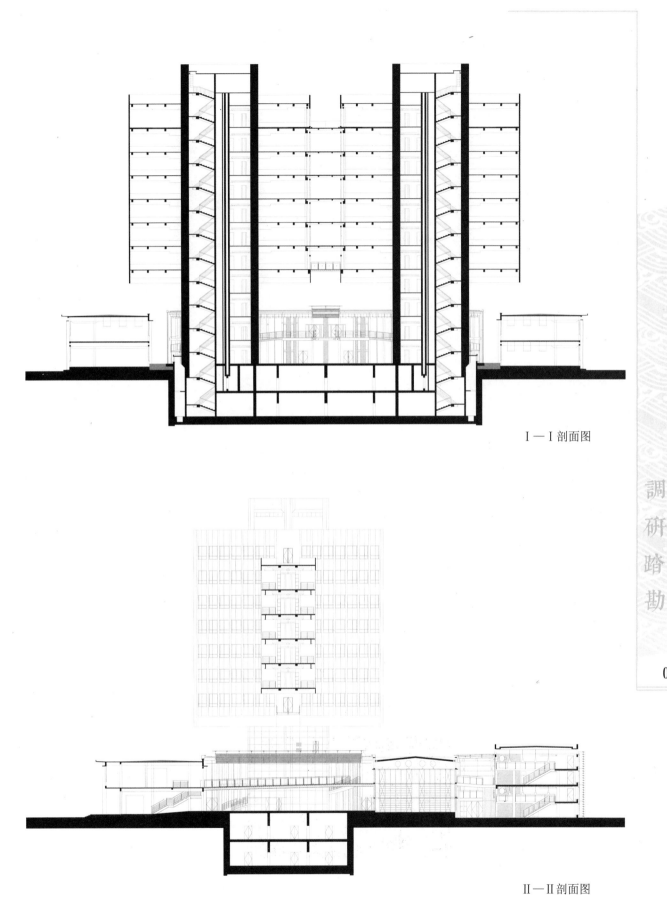

Ⅰ—Ⅰ剖面图

Ⅱ—Ⅱ剖面图

國粹賡續卷——歷史建築保護與再利用設計

調研踏勘

华南理工大学
South China University of Technology

华工附属中学旧建筑再生——高校众创空间设计
Regeneration of an Old Building in the Affiliated Middle School of South China University of Technology—Intramural Public Entrepreneurial Space Design

小组成员：汪瀚　杨明之　邓康利

项目背景 | Project Background

本案选址于华南理工大学五山校区北区一组有着近七十年历史的建筑群，设计目标是将其重新改造利用，形成校内一处大学生创新创业孵化基地（简称众创空间）以满足大众创业、万众创新背景下本校学生日益增长的创业孵化需求。

基地总占地面积约 12500 m²，毗邻华工五山校区主干道长江北路与北区中心湖。场地中一栋三层，二栋两层的砖混结构建筑由单层连廊沟通，总建筑面积约 8400 m²。该建筑群建成于 1952 年，是华工独立办学后首批建设的大学配套设施之一，最初作为大学附属中学的教学用房。

国立中山大学工学院 1949　华南工学院 1952　华南理工大学 1988

南北区功能对比图　南北区学生活动热度图

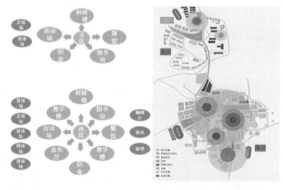

华南理工大学五山校区分南北两区。南区由国立中山大学时期老校园发展而来，北校区则始建于华工独立建校之后。相比而言，南区主要承担教学、科研、行政等综合功能，而北区主要承担师生生活配套功能。相比而言，北区功能类型单一、缺乏活力。

基地周边功能　基地周边交通

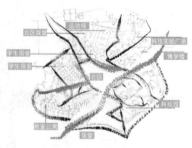

基地位于华工五山北校区中心湖东侧的台地之上，与学校实验用房、宿舍区比邻。

北区有两条主要交通流线，北湖南路与长江北路。项目基地位于"长江北路"东侧，便捷可达，但识别性较差。

基地流线与植物分布图

原有基地交通流线简单,以中央圆型花坛组织,形成环形流线,主入口位于基地西侧靠近长江北路。基地原生植被丰富,古木参天,有数量众多的木棉、白千层、榕树等高大乔木。

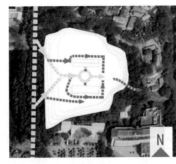 

建筑分布示意图

一号楼建模初步建模图

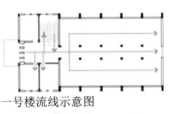

一号楼流线示意图

建筑群东侧三层中央主楼(原1号楼)建筑面积约1200 m²,框架结构,内部空间开放灵活。

二号楼建模初步建模图

二号楼内部现状图

建筑群南北两侧分别为二号、三号楼,砖混结构,图中红色部分表示为不可改动的承重墙、柱。

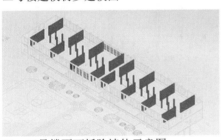

二、三号楼不可拆除墙体示意图

二号楼走廊现状图

因为建筑最初作为中学教室使用,室内空间呈鱼骨式均匀分布。二号楼二层走廊上方的高窗,为室内带来自然通风与采光。

建筑群东侧三层中央主楼(原1号楼)建筑面积约1200 m²,框架结构,内部空间开放灵活。

优劣势分析

优 与华工同岁,功能几经波折,充满故事,外立面充满了50年代华工建筑的特色。

劣 建筑局部维护结构老化、局部屋面漏水现象严重。原有建筑门窗老旧,且与现有建筑节能要求不符。

优 地处北区中部,基地面朝湖泊并被古树围绕,处在一个相对独立的空间内,不易受到外界干扰。

劣 庭院内部缺乏公共空间,导致在看似一个体系的庭院中的三个建筑的使用者之间没有交流和互动。

初步设计理念

创建高校中的创新创业社区

附件：建筑与场地图

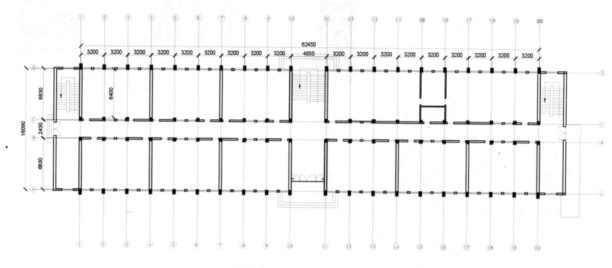

二号楼首层原始平面图

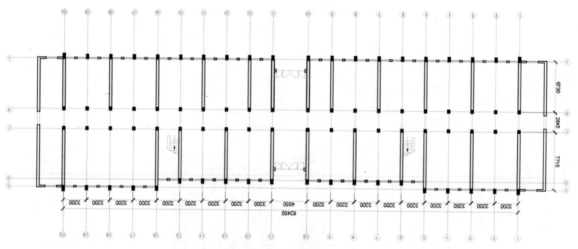

二号楼二层原始平面图

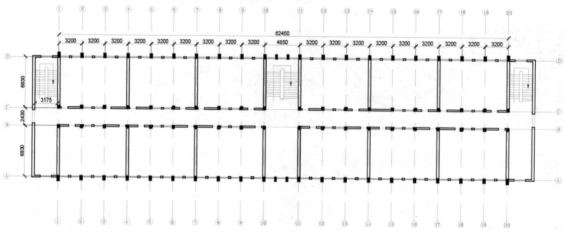

三号楼首层原始平面图

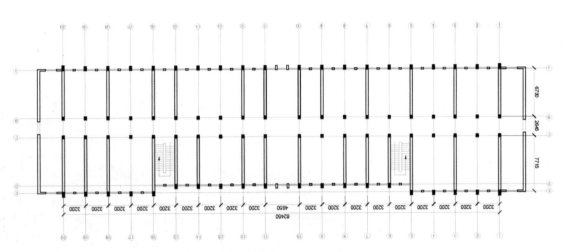

三号楼二层原始平面图

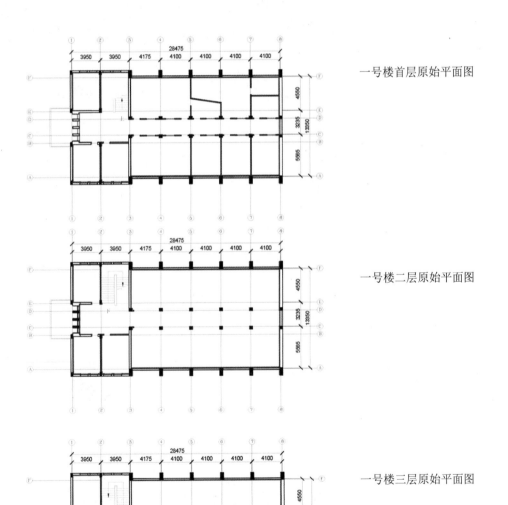

一号楼首层原始平面图

一号楼二层原始平面图

一号楼三层原始平面图

华南理工大学
South China University of Technology

30个模块+7个技术——云县传统民居更新设计
30 Modules+7 Technologies—Renewal Design of the Traditional Dwellings in Yun County

「室内设计 6+」2018（第六届）联合毕业设计
"Interior Design 6+"2018(Sixth Year) Joint Graduation Project Event

小组成员
王　铭
张豪元
王璐瑶

项目背景 | Project Background

本案中国村庄位于云南省临沧市云县昔宜村，古村原状采用乡土材料建设，充满地域特色，周边旅游资源丰富。现在村民面对生活质量提升，又有旅游方面的需求。有政府指导的新居方案，但不适应农民生活，村民自己新建的新居失去古村特色。

- 昔宜村
- 国道
- 白莺村
- 白莺山

交通环境分析

昔宜村建筑群落航拍模型图

调研发现 | Research Findings

调研传统特色总结（建筑类型）

- 四合院
- 三合院
- 长屋

调研发现 | Research Findings

以昔宜村罗良海家旧宅四合院为完整模板的传统建筑功能布局调研分析

利用山坡地形分为地面一层和地下负一层，干栏式建筑，地面一层传统用作会客空间，故称为"面子楼"，地下一层用来饲养牲口。

厨房（侧楼）

一般紧邻主楼，作为家庭食物加工空间，包括制作腊肉等食材

主楼

一般分为两层两进三开间，一层为主人起居功能，中间一厅两侧为卧室，二楼多用来储藏粮食。

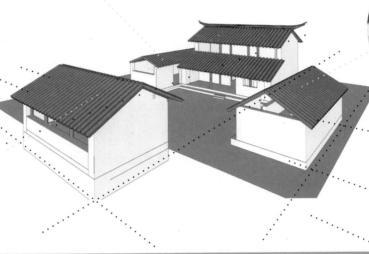

面楼

杂物间（副楼）

一般位于主楼右侧，多用来堆放干柴杂草和农具，有些作为卧室使用。

其他传统特色

火塘　　叉叉锅　　在墙上的鸡窝

①分家另立火塘，标志着家庭的分化，由一个家庭中分化出的血缘关系的家庭便渐渐形成了一个家族。
②在某些民族的家庭中，火塘分别有男火塘和女火塘，于是火塘又具有性别的象征。

通过自由调节高度来调节温度的锅，极具当地智慧，是火塘的好搭档。

土墙配鸡笼，给鸡准备的"窑洞"

问题剖析 | Problem Analysis

新建建筑问题

1. 对于政府统一规划的新屋，如何保留传统建筑形式，政府没有给出标准范例的图纸。

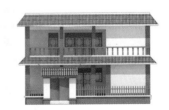

政府统一规划下的新建民居

國粹賡續卷——歷史建築保護與再利用設計

調研踏勘

2. 前面提到的传统特色，在平面图中几乎消失殆尽。四合院也好，灶台也好，火塘也好，墙上的鸡笼也好，统统成为了为满足现代都市生活的建筑。

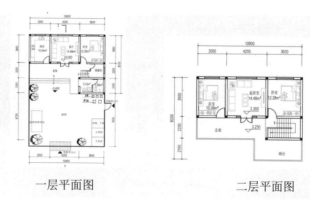

一层平面图　　　　二层平面图

3. 对于村民自建新居，村民在自建的过程中往往也没完全按政府统一规划的要求来建设。当地传统建筑明显特征被全部略去，建筑材料由原本的木结构转变为清一色的砖混结构，建筑的内外都发生了很大改变。

古建筑残留问题

厕所卫生条件简陋　主要在猪圈等牲畜棚里解决　　清洁条件简陋　洗澡主要错开男女，在澜沧江上洗浴　　老屋瓦片漏雨　屋顶经常会出现漏雨现象

问题剖析（村民的真正诉求） | Problem Analysis

 政府　+　 设计师的思考　+　 村民

"要政绩"
"要开发旅游资源"
"要体现地域建筑特色"

"如果无条件呼吁保护老房子，对当地村民也是一种伤害。"

"但看着传统建筑特色在一张设计图后成为了千村一面中的另一个小村，非常可惜。"

只想要"更好的生活条件"，哪怕仅仅满足现代都市功能的建筑对于当地村民也非常不错。

解决这些问题不是具体的设计方案，而是需要一个设计导则

工作思路 Working Ideas

问题	目标	策略	技术手段
1个问题	3个目标	19个建筑模块 11个功能模块	7个技术
千村一面 如何留住乡愁	改善生活 文化传承 旅游需求	生活 住宿…	抗震改造 消防方式…

三点设计任务目标

改善居民生活
即通过改善居民居住环境,增加新功能。

传统文化传承
通过分析主要问题,本设计的主要任务为将功能模块化,倡导在外观上还是保留老建筑特征,尤其是保留了老建筑的布局,同时在内部也能有所变化,加入现代生活的必要功能,保留一部分对于他们生活来说重要的传统功能。

满足旅游发展需求
满足由旅游业发展带来的大量城市游客的需求。

像拼积木一样的模块化处理

为什么选择进行模块化设计(模块化的优势)

解决"千村一面"
和政府大包大揽的设计策略不同,让每一家住户通过不同的模块组合,使自己的房屋都有独有的特点。

旅游等新型需求的介入

村民缺少规划功能的能力

功能选择权交给村民
选择传统功能还是选择城市现代民宿功能,这个权利还是要由真正的建筑使用者来决定。

政府需求
政府需要一个可以满足大部分房屋旧改的导则。

附件:建筑与场地图

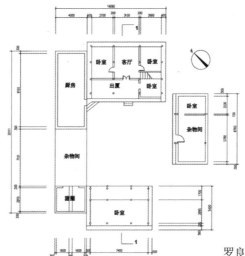

罗良海旧宅四合院平面图

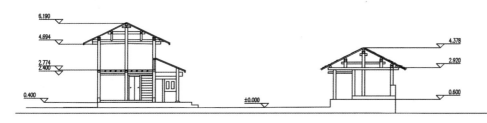

罗良海旧宅四合院剖面图

哈尔滨工业大学
Harbin Institute of Technology

文化寄居——哈尔滨道里文化宫空间改造设计
Cultural Sojourn——Harbin Daoli Cultural Palace Spatial Reconstruction Design

小组成员
陈博文
韩　秋
张宇淳
朱雪莹

基地概况 | Base Overview

区位分析 | Location Analysis

■ 一类保护建筑
■ 二类保护建筑
■ 三类保护建筑

基地位于哈尔滨道里历史城区，老建筑密集，临近中央大街，历史氛围浓厚，但目前缺少合适有效的开发和利用。

基地周边著名景区较多，但联系不够密切。因此，基地本身需提升自身吸引力以此提升旅游价值和经济价值。

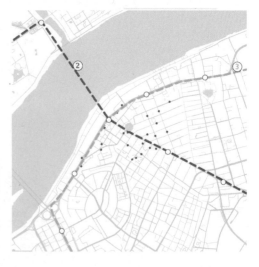

基地周边公交车站众多。人民广场站为2号线与3号线换乘车站，基地临近人民广场站，交通便利，发展潜力大。

业态分析 | Business Type Analysis

- 文化宫
- 展览馆
- 剧院
- 手工艺品商店
- 书店
- 电影院
- 老年活动中心

建筑年龄分析 | Building Age Analysis

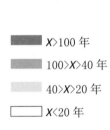
- $X>100$ 年
- $100>X>40$ 年
- $40>X>20$ 年
- $X<20$ 年

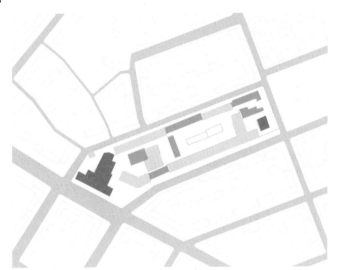

人口构成分析 | Population Composition Analysis

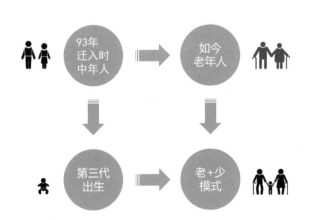

居住条件较好，距广场、江边较便捷，缺少小区内部的活动场所。

附件：建筑与场地图

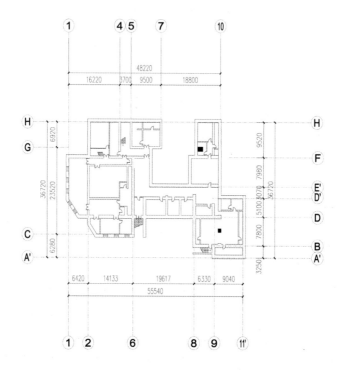

文化宫一层平面图

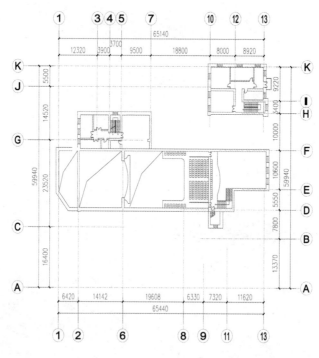

文化宫二层平面图

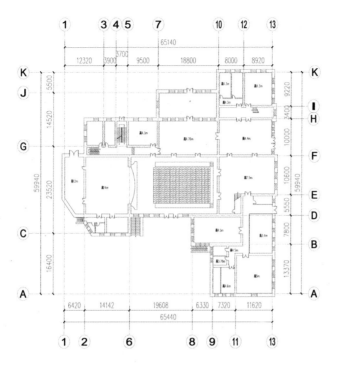

文化宫地下一层平面图

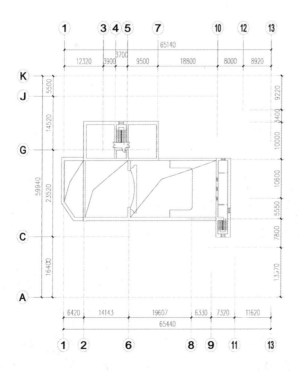

文化宫三层平面图

哈尔滨工业大学
Harbin Institute of Technology

历史建筑保护背景下建筑空间及其社区环境改造设计
Renovation Design of an Architectural Space and its Surroundings in the Context of Historic Building Conservation

「室内设计 6+」2018（第六届）联合毕业设计
"Interior Design 6+"2018(Sixth Year) Joint Graduation Project Event

小组成员
周子钦 颜岩 赵斌 周毓

基地概况 | Base Overview

黑龙江省省会哈尔滨，简称"哈"，别称"冰城"，是副省级市、特大城市、中国东北地区中心城市之一，哈尔滨都市圈核心城市，是东北北部交通、政治、经济、文化、金融中心，也是中国省辖市中陆地管辖面积最大、户籍人口居第三位的特大城市，地处中国东北平原东北部地区、黑龙江省南部。

道里区是哈尔滨市的中心区，位于北纬45°32′—47′，东经126°08′—38′之间，全区总面积517.2km²，其中市区面积22.6km²。道里区境内松花江自西向东经过。

基地被红霞街、高谊街、中医街和经纬街包围，临近红星公园、中央大街。哈尔滨属中温带大陆性季风气候。2007年，道里区政府对这座哈尔滨市一类保护建筑进行了保护性修缮，现更名为道里区文化宫，也是道里区文体局的所在地。

区位分析 | Location Analysis

交通：基地周边公交车站众多。人民广场站为2号线与3号线换乘车站，基地临近人民广场站，交通便利，发展潜力大。

历史保护建筑：基地位于哈尔滨道里历史城区，老建筑密集，临近中央大街，历史氛围浓厚，但目前缺少合适有效的开发和利用。

绿化：基地所在道里区为哈尔滨老城区，距离面向城市市民的绿化活动空间较近，但面向社区居民的绿化活动空间除红星广场外较少，有待增加。

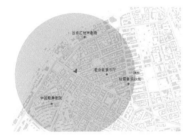

基地周边一公里内音乐厅分布

基地周边一公里内展览馆分布

基地周边一公里内艺术教育分布

建筑空间构想 | Architectural Space Conception

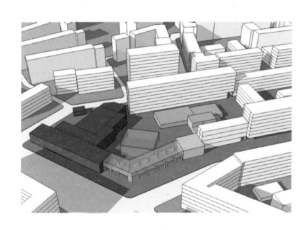

对老建筑及新建筑进行相关的室内设计，对老建筑的功能、形式、空间的丰富程度进行新的补充。从而带动整个社区乃至周边社区的发展。

空间的重塑——补全街区立面（加建）公共空间的营造、新老建筑的立面连接、最大程度利用城市立面。

空间的打破——引入城市轴线，与周边街区建立联系。

附件：建筑与场地图

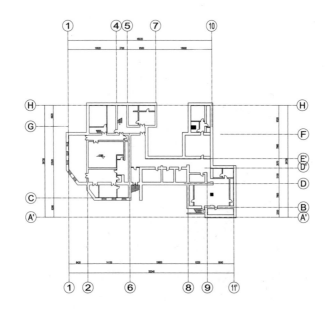

地下一层平面图

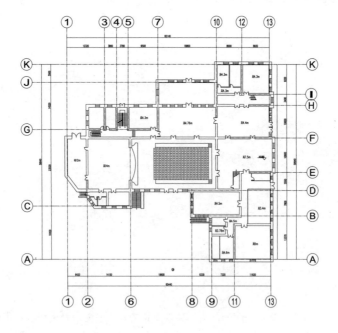

一层平面图

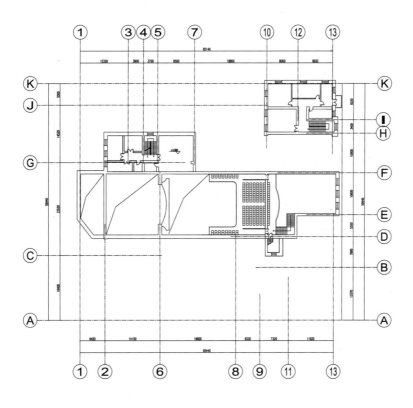

二层平面图

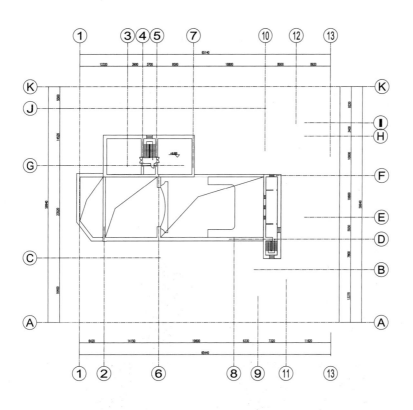

三层平面图

西安建筑科技大学
Xi'an University of Architecture and Technology

山西省夏门村梁氏古堡建筑群修复与再利用设计

Renovation and Reuse Design of Liang Family Castle Building Complex in Xiamen Village, Shanxi Province

「室内设计 6+」2018（第六届）联合毕业设计

"Interior Design 6+"2018(Sixth Year) Joint Graduation Project Event

056

小组成员
崔维鹏
於天心
马永强
周文婷

基地概况 | Base Overview

区位 | Location

夏门村位于山西省晋中市灵石县境内，在其周边则有介休、汾阳以及古城平遥与之相邻，使夏门村有着较为优越的区域文化资源。

夏门村背倚吕梁山脉的秦王陵，前瞻太岳山脉的韩信岭，三晋母亲河、黄河第二大支流——汾河，在脚下流淌，龙头岗在堡后耸立。形成了前对削壁以为屏，后倚峻岭以为靠，下临汾水以为险，底坐磐石以为基村落格局。

夏门村原是隋唐时期的古战场，后因军事地位下降、商业发展以及梁氏家族的迁入逐渐形成村落。由于夏门村土地资源稀缺，村落建筑依山而建，在晋商文化的影响下，形成了当地特有的"跃层式堡寨窑洞"，而"梁氏古堡建筑群"则是其独有的特色代表。

村落 | Village

夏门村主要由梁氏古堡建筑群及村落东侧民居构成，新建建筑居于主干道路南侧。当地产业类型主要以水泥厂（仍在使用）、煤矿场（现已关闭）、洗煤厂（现已关闭）为主，分布在村落外围。种植绿地较少主要沿汾河河岸呈带状分布。村落建筑以梁氏古堡建筑为代表，古堡建筑因清朝时期梁氏族人高中进士而建立，距今已有上百年。

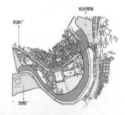

古堡 | Castle

夏门村主要建筑包括御史院、大夫第、永宁堡、知县宅、古民居等。而梁氏古堡则包括了御史院、大夫第、永宁堡及知府院四个建筑群落，为此，又称其为"夏门古堡"。

梁氏古堡因"筑堡卫家"而建造，所以从建造之初，防御的思想便深入其中，作为"住守合一"的建筑体系，可居、可匿、可防、可退，在古堡建筑中各个建筑群之间有暗道相连用于防卫，从而也形成了"立体交叠、明暗互通"的立体交通。

梁氏古堡建筑群位于夏门村落最东侧，此处地势最高、地形坡度最大，依壁而建的百尺楼则成为这里最先抵御外敌的哨塔。

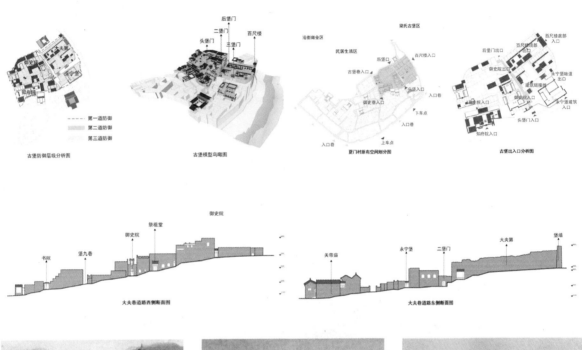

总结 | Conclusion

优势 | Strengths

1. 历史文化悠久
2. 交通便利
3. 建筑类型独特
4. 文化传统丰富

问题 | Problems

1. 古堡建筑群年久失修，出现破损
2. 年轻人口外迁，文化传承出现断层
3. 建筑空间利用较差，功能缺失
4. 人居生活环境较差、环境污染较重

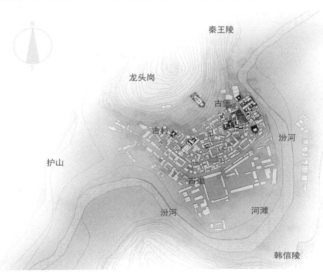

附件：建筑与场地图

夏门村村落平面图

夏门村布局平面图

夏门村建筑布局平面图

永宁堡平面图

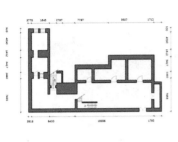

一层平面图

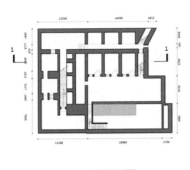

二层平面图

三层平面图

大夫第平面图

百尺楼平面图

一层平面图　二层平面图　三层平面图　四层平面图

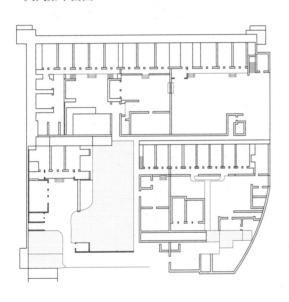

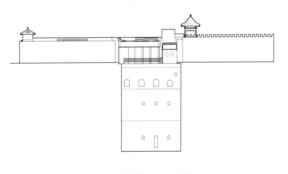

百尺楼东立面图

御史院平面图

知府院平面图

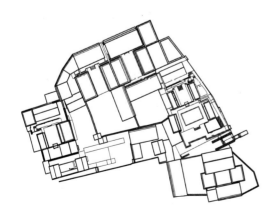

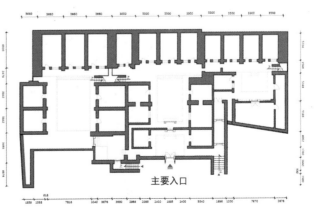

主要入口

西安建筑科技大学
Xi'an University of Architecture and Technology

韩城古城的保护与再利用设计
Conservation and Reuse Design of Hancheng Ancient City

「室内设计 6+」2018（第六届）联合毕业设计
"Interior Design 6+"2018(Sixth Year) Joint Graduation Project Event

小组成员
张景一
王玲子
任晓贤
曹玥玲

基地概况 | Base Overview

韩城古称"龙门""夏阳""少梁"，是中华文明重要发祥地之一，也是史圣、太史公司马迁的故乡。

韩城位于陕西省东部黄河西岸，西河重镇，川原怀抱。南临居水，西依梁山，东北有塬。在地势上呈现出"西北高、东南低"的特征。

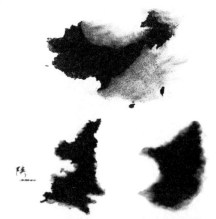

韩城卫星图

建筑现状 | Existing Building Conditions

发现问题 | Problem Finding

1. 古城建筑年久失修，出现坍塌破损的状况。
2. 手艺人的大量搬迁导致非物质文化遗产的流失。
3. 古城居民缺乏活动空间。
4. 四合院内功能不完善，缺乏厨房、卫生间。
5. 居住设施相对较差，如用水、用电、排水、采暖等，不能完全满足生活需求

预期处理办法 | Expected Treatment Method

保留尺度比例

加强修缮

按照传统风格新建

改造完善厨卫用水

减少私拉隐患

改善冬季室内环境

附件：建筑与场地图

基地模型

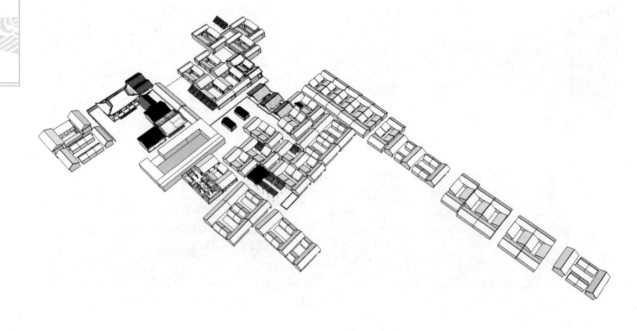

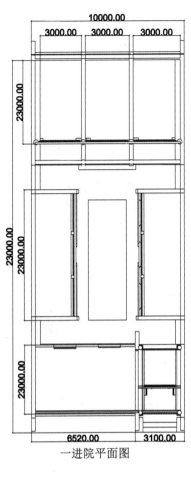

一进院平面图

二进院平面图

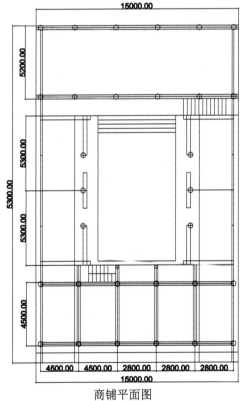

商铺平面图

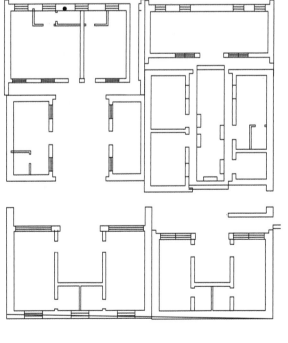

民宿平面图

北京建筑大学
Beijing University of Civil Engineering and Architecture

恭王府博物馆展览设计
Prince Kung's Palace Museum — Exhibition Design

「室内设计 6+」2018（第六届）联合毕业设计
"Interior Design 6+"2018(Sixth Year) Joint Graduation Project Event

小组成员
刘宇麒　韩　玥
闫　铮　杨玉萍
曹志玮　郑一霖

基地概况 | Base Overview

恭王府博物馆——
国家级非物质文化遗产保护与展陈基地

恭王府位于北京西城区什刹海西岸的前海西街，是目前北京保存最完整且唯一对社会开放的清代王府建筑群。近年来，恭王府博物馆作为国家非物质文化遗产保护与展陈基地，以"中华传统技艺系列精品展"为工作载体，紧紧围绕中国非物质文化遗产开展保护系列活动。

1982年被列入第二批全国重点文物保护单位。恭王府始建于清乾隆四十一年（1776年），至宣统退位，恭王府共历经了7位皇帝的统治时期，见证了清王朝由胜至衰的整个历史过程，故有"一座恭王府，半部清代史"之说。1988年，恭王府花园对外开放，2008年恭王府完成府邸修缮工程后，全面对外开放。

忻州秀容书院——
国家级非物质文化遗产保护与研究工作站

忻州古城始建于东汉建安二十年（公元215年）距今已有1800多年历史，地处中原农耕文明和北方游牧文明结合部，地形险要，位置突出，故历代多为郡、州治所，常为商贾往来、兵家必争之地，素有"三关总要""晋北锁钥"之称。

忻州工作站位于山西忻州，是由恭王府国家级非遗展陈基地与忻州市政府共建。忻州老城是一个具有浓郁地方特色的历史文化保护片区，总用地面积192.38hm²。

规划方案以发展的观念保护旧城，实现将历史文化资源转变为产业目标。

调研情况 | Research Results

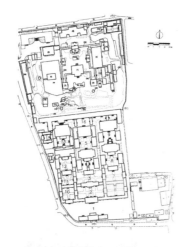  

王府整体由府邸和花园两部分组成，南北长约330m，东西宽约180m，整体略呈一个倒梯形，总占地面积6万m²。府邸部分平面布局为"三路五进式布局"，分为东、中、西三路建筑，每路皆由多进四合院组成。

展厅是位于王府东二区院落，东、西倒座房皆为五开间，一进深；东、西厢房皆为三开间，一进深。

优势
1. 恭王府作为国家级非物质文化遗产保护与展陈基地，建筑本身独具特色。
2. 基地花园提供了活动场地。
3. 王府内有许多常设展览，文化氛围浓厚。

劣势
1. 房间开口较多，内部行走路线混乱。
2. 古建筑自身条件远达不到现代展陈要求，进深小无法达到观展照相距离。
3. 倒座房内有之前展览遗留下的展柜，院落中间还有上一个展览留下的临时建筑。
4. 展厅内为符合历史建筑保护要求，加盖了天花、射灯及空调装置，墙壁上安装展板。

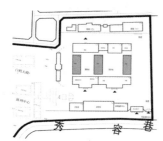

秀容书院是山西省重点文物保护单位、唯一保存完好且具有教育功能的书院，始建于清乾隆四十年（1775年），忻州知州鲁潢（江西新城人）倡导创造秀容书院，选址城西南九龙岗头的白鹤观。

秀容书院位于忻州市忻府区旧城西南角，坐北向南。当时忻县称秀容县，故以此得名，为忻州市第一所学府。秀容书院依地形而建，由上、中、下三院组成。上院为主院，中、下院为书舍（这三院也是我们的展厅位置）。每个书舍面积约为78 m²。

展览内容介绍 | Information of Display Content

杨虾与同事讨论设计家具

广州十三街洋行

山东杨泉村铁木加工厂

两个不同地域文化差异造就了两种截然不同的广作与晋作家具文化。

晋作家具位于中国中部地区，浑厚凝重深受当地人们喜爱，家具制作技艺主要以家庭传承的方式沿传。晋作家具展览以家庭传承方式与背景、木材用料、制作技艺、经典纹样四方面进行展览设计。

广作家具位于广东、广西一带，受西方文化影响较大，尤其以巴洛克、洛可可风格为主。广作家具展览以历史谱系、制作技艺、经典纹样、传承人杨虾家族四方面进行展览设计。

南京艺术学院
Nanjing University of the Arts

万花筒——民国时期人物展
Kaleidoscope—Personages of the Republican Period

『室内设计 6+』2018（第六届）联合毕业设计
"Interior Design 6+"2018(Sixth Year) Joint Graduation Project Event

小组成员：马悠庭 范聪达 高金宇 孙昱

基地概况 | Base Overview

南京总统府位于南京市玄武区长江路292号，是中国近代建筑遗存中规模最大、保存最完整的建筑群。占地面积约为5万余㎡，既有中国古代传统的江南园林，也有近代西风东渐时期的建筑遗存，至今已有600多年的历史。1912年1月1日，孙中山在此宣誓就职中华民国临时大总统，辟为大总统府，后来又为南京国民政府总统府。

建筑族群分析 | Building Complex Analysis

  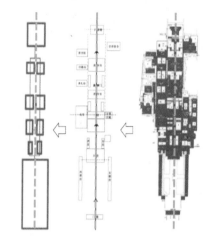

南京总统府中轴线建筑组群的组合是递进结构。一方面表现在建筑功能要求上，另一方面表现在空间的情感体验上。总统府内中轴线上现存的建筑而言，混合了从清末到民国多个时期的多种类型建筑样式，通过轴线把不同的单元空间组织成一个可连续感知的整体。

装饰形式 | Decorative Form

拱　形　　拱形门、拱形落地窗、柱式

地面：青砖、耐火砖片、水泥石粉、水刷石面砖

方　形　　方形空格窗、棱形门

墙面：青砖、耐火砖片、水泥石粉、水刷石面砖

曲线形　　回形纹样、卷涡纹样、植物纹

结构：砖木结构、钢筋混泥土结构

装饰质感、肌理、材料

装饰色彩分析 | Decorative Color Analysis

灰色：大堂、二堂的青瓦屋顶、接待室外墙群

红色：大堂的立柱、梁架、礼堂通道外立面、政务局红瓦屋顶

黄色：接待室外立面、子超楼外立面

界面装饰提取 | Interface Decoration Extraction

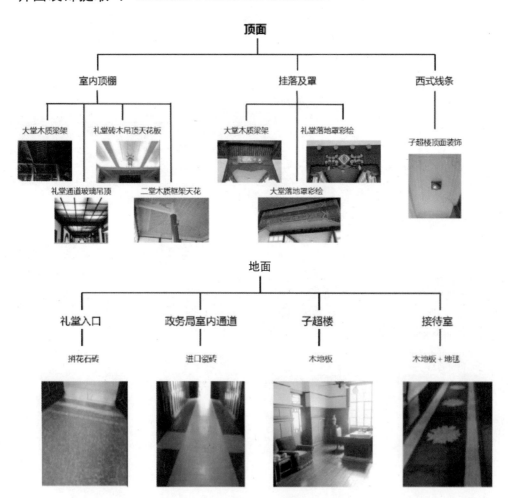

附件：建筑与场地图

总统府平面图

目标建筑分布图

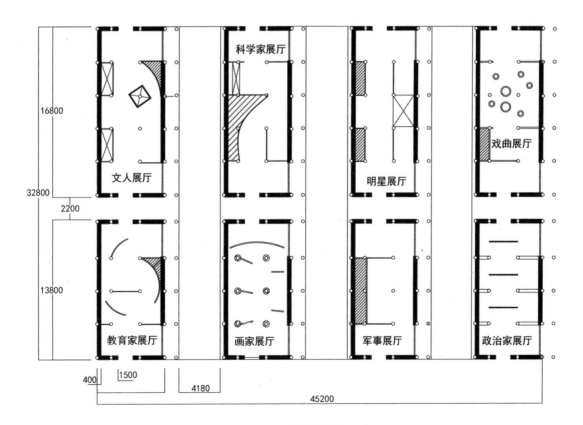

展厅平面图

浙江工业大学
Zhejiang University of Technology

合游·浙江省薛下庄村"二十四间"老建筑保护与再利用设计

Heyou—Conservation and Reuse Design of "Twenty-four" Old Buildings in Xuexia Village, Zhejiang Province

小组成员 章家骐 章佳祺 刘叶 蒋一德

区域区位及概况 | Location and Overview

薛下庄地处斗鸡岩下，壶源江流绕村蜿蜒而过，依山傍水，风光秀丽。在薛下庄村以薛姓居多，自称莲塘薛氏，与下薛宅薛姓人本系同宗。早在元朝初年，薛氏先祖松堂公之十五世孙桂吉公便从下薛宅移居至此，创基筑室，繁衍生息，迄今已历二十四世，逾七百载。薛下庄又称薛家庄或薛家村。

村子虽紧贴壶源江，因水流相对较浅，向无舟楫之利。旧时，村民与外界的联系全赖上下两段过江蹬步。遇洪水泛滥，蹬步被淹没，斗鸡岩下的盘山小路就是出村的唯一通道。眼下，环村二古二新四座石桥并存，凸显水乡特色。最小的一座古桥在村南、江之彼岸，名"保元桥"，建于清光绪己卯（1879年），单孔，条石拱砌，呈八字形，为旧时金坑古道所必经。

基地分析 | Base Analysis

木结构建筑，建造时间最早，质量整体较差。

夯土建筑，建造于地面20世纪中期，保护价值较低。

砖砌建筑，建造于近30年，质量好，无历史价值。

有机更新理论

早期建筑
20世纪八九十年代建筑
新式建筑

三种建筑
- 有文化意义的建筑
- 质量差构成危房的建筑
- 保留与拆除之间的建筑

经济社会结构

城市中所谓的"衰败地区"，由于地区物质环境的衰败导致地方税收减少和市政补贴的增加。做好城市更新，有助于提高地区的经济活力，复苏竞技，增加城市的繁荣。

城中村改造

概念阐述 | Concept Description

"旅游+互联网+农业+创客"的模式

1. 在古老的木材结构下，进行着新兴的工作与研究。
2. 旧与新，传统与科技，过去与未来，对比鲜明，却又交相辉映。
3. 通过对闲置农业生产设施的改造，植入新的业态。
4. 与空间改造同步，一系列与古村相关的文创产品和旅游活动内容也被一起考虑。

| 旧 | 古老的木材 | 传统 | 过去 |
| 新 | 新兴的工作 | 科技 | 未来 |

诉求 | Demand

原住民A

希望能够继续住在"二十四间"中，毕竟已经在这里住了一辈子了。

原住民B

"二十四间"是我们这一代人的记忆，不能够将它拆除，最好能给村里带来新的生气。

官员C

村里需要发展旅游业，"二十四间"是村里面最好的一个资源点。

学者D

"二十四间"建筑是一幢富有历史气息的建筑，它的保护和再利用对于建筑乃至文化都是至关重要的。

附件：建筑与场地图

场地总平面图

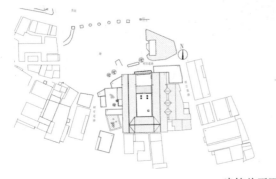

建筑总平面

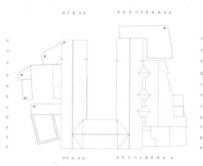

屋顶平面图

一层平面图

二层平面图

北立面图

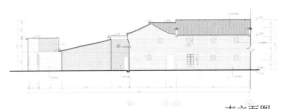

南立面图

东立面图

西立面图

1-1 剖面图

2-2 剖面图

3-3 剖面图

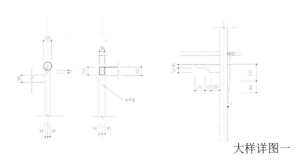

大样详图一

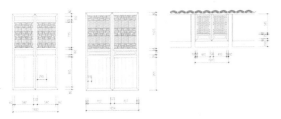

大样详图二

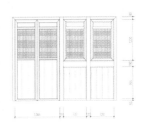

大样详图三

热点命题,纷显特色

联合指导,服务需求

中期检查

室内设计6+1 2016（第六届）
联合毕业设计
Interior Design 6+ 2018 (Sixth Year)
University Enterprise Cooperative Graduation Project Event

同济大学
Tongji University

同济大学图书馆室内外环境保护与再生设计
Conservation and Regeneration Design of the Indoor and Outdoor Environment of Tongji University Library

「室内设计 6+」2018（第六届）联合毕业设计
"Interior Design 6+"2018(Sixth Year) Joint Graduation Project Event

小组成员
李一丹　马一茗　潘蕾
张迪凡　刘雨婷　张奕晨
许可　　毛燕　　DANA

概念提出 | Bringing Forward Conception

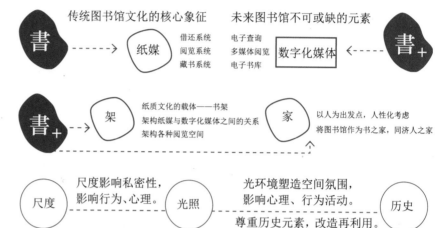

尺度分析 | Scale Analysis

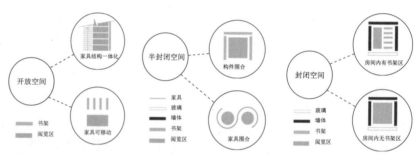

开放空间主要适用于大面积的阅览；半封闭空间隔离视线，围合构件有多种选择，可用于阅览、自习、讨论等；封闭空间隔离声音，主要用于研习室、影音室。

光环境分析 | Light Environment Analysis

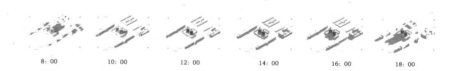

8:00　10:00　12:00　14:00　16:00　18:00

图书馆整体设计思路 | Overall Idea of the Library Design

流线 | Flow Line

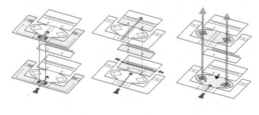

裙房：对内开放角部楼梯间。

书库：不更改流线，对书库的室内通行空间进行改善，增加标识性。

塔楼：不更改流线，对大厅核心筒入口进行引导的改善。

功能植入 | Function Implantation

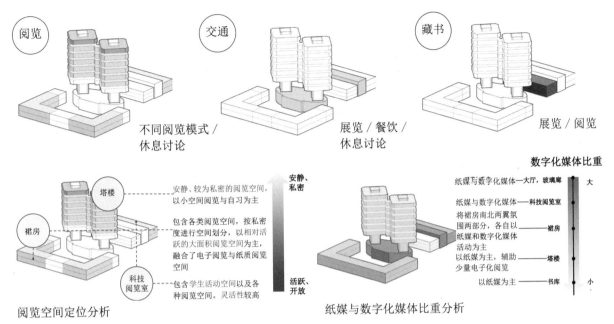

阅览 不同阅览模式／休息讨论

交通 展览／餐饮／休息讨论

藏书 展览／阅览

塔楼：安静、较为私密的阅览空间，以小空间阅览与自习为主

裙房：包含各类阅览空间，按私密度进行空间划分，以相对活跃的大面积阅览空间为主，融合了电子阅览与纸质阅览空间

科技阅览室：包含学生活动空间以及各种阅览空间。灵活性较高

阅览空间定位分析

安静、私密 — 活跃、开放

数字化媒体比重

纸媒与數字化媒体—大厅，玻璃廊

纸媒与数字化媒体—科技阅览室

将裙房南北两翼氛围两部分，各自以纸媒和数字化媒体活动为主 — 裙房

以纸媒为主，辅助少量电子化阅览 — 塔楼

以纸媒为主 — 书库

大 — 小

纸媒与数字化媒体比重分析

方案设计 | Conceptual Design

主要设计部分为裙房、大厅、书库和塔楼标准层。裙房定位为活跃的综合阅览空间，大厅定位为多活动融合空间，书库定位为展览和阅览结合的体验空间，塔楼定位为安静的综合阅览空间。

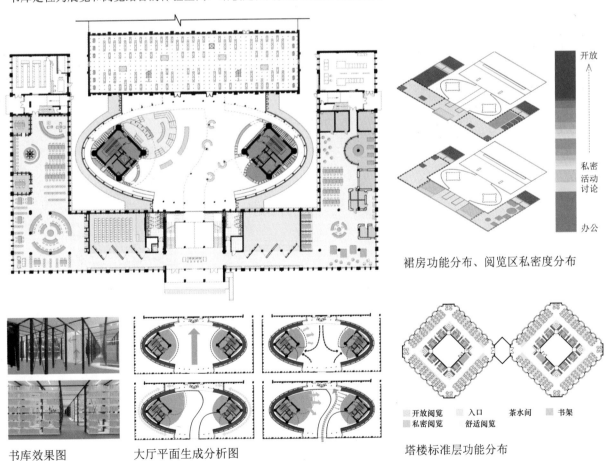

裙房功能分布、阅览区私密度分布

开放 — 私密 — 活动 — 讨论 — 办公

书库效果图

大厅平面生成分析图

塔楼标准层功能分布

■ 开放阅览　■ 入口　　　■ 茶水间　■ 书架
■ 私密阅览　■ 舒适阅览

华南理工大学
South China University of Technology

华工附属中学旧建筑再生——高校众创空间设计
Regeneration of an Old Building in the Affiliated Middle School of South China University of Technology—Intramural Public Entrepreneurial Space Design

小组成员 汪瀚 杨明之 邓康利

提出任务书 | Assignment Planning

- **A** 众创空间实地调研
  - 华南理工大学校内现有众创空间实地调研
  - 广州工业大学、广州大学等高校校内现有众创空间实地调研
- **B** 校外众创空间调研
  - 深圳华强北创客空间
  - 前海青年创客空间
- **C** 国内外经典案例参考
  - 北京远洋新干线、we+联合办公空间
  - 芬兰 "DESIGN FACTORY"
- **D** 深入访谈
  - 对创客群体的访谈
  - 对创业管理与运营团队的访谈
- **E** 问卷调查
  - 建成使用后评价（POE）
  - 功能使用需求调查
  - 展示推广需求调查

通过问卷调查，我们收到有效问卷166份，归纳起来我们发现：
1. 创客们对咖啡厅、书吧、健身房和自助超市需求较高。
2. 创业管理者对创新创业孵化基地的服务类功能较为需要。

功能布局	对应	一号楼	二号楼	三号楼	室外
办公空间	独立办公空间		√	√	
	联合办公空间		√	√	
	老师管理室	√			
	学生管理室		√	√	
	接待空间	√	√	√	
	服务大厅	√			
展示区	成就展示区	√			
	VR展示区	√			
	作品展示区	√			√
会议空间	行政会议室	√			
	团队会议室		√	√	
	公共会议室		√	√	
	沙龙区	√			
	老师培训区		√		
休息空间	阅读自习区	√		√	√
	咖啡区		√	√	
	睡眠区		√	√	
	座位区	√	√		
	健身区		√		
	厕所区	√	√	√	
制作空间	打印区	√	√		
	中大型机械区			√	
储藏空间	储藏区	√	√	√	

功能任务书

辐射北区

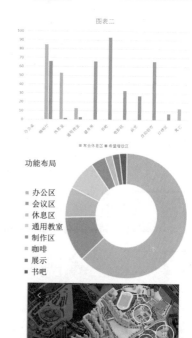

总平面图

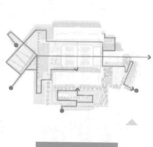

- 苗圃区
- 咖啡区
- 东入口
- 冥想区
- 运动区
- 广场区
- 滑板区

基地流线图

设计深化 | Design Detailing

通过访谈和问卷调查，我们发现高校创业人员普遍存在对运动的爱好和需求。"一起运动"是联系创客群体以及校园不同人群之间最自然的纽带。因此，我们将室外场地融入多样的运动功能，既满足创客的运动需要，也形成大学校园中定期举办创客沙龙的集会场所，提升校园中人与人之间的碰撞交流的可能性，使该基地最终成为校园中的青春洋溢、活力四射的创意创新社区。

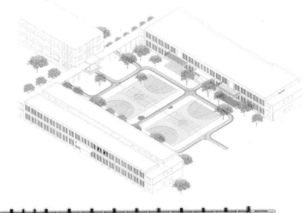

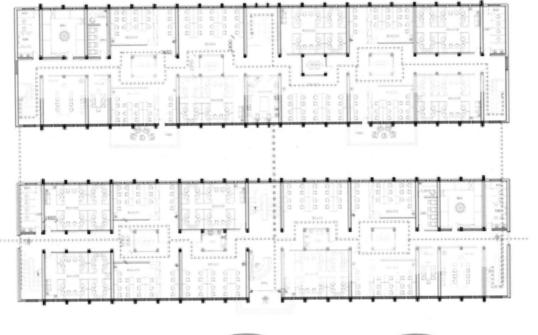

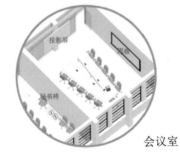

会议室

制作间

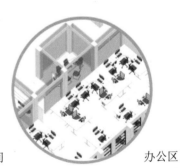

办公区

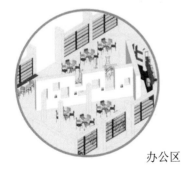

办公区

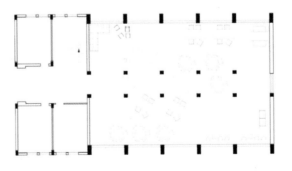

华南理工大学
South China University of Technology

30个模块+7个技术——云县传统民居更新设计
30 Modules+7 Technologies—Renewal Design of the Traditional Dwellings in Yun County

「室内设计 6+」2018（第六届）联合毕业设计
"Interior Design 6+"2018(Sixth Year) Joint Graduation Project Event

小组成员：王铭　张豪元　王璐瑶

设计导则 | Design Guidelines

原传统建筑的主要四大功能与多项辅助功能

主要功能：
- I 前厅/前台+卧室+厕所
- II 灶房/厨房+餐厅
- III 生产间+杂物间
- IV 活动间+火塘间……

+

辅助功能：养殖、小展厅、自助超市、小酒吧、儿童游戏、公共洗衣服……

将传统建筑进行模块划分

一层平面（四合院）　（主楼）（副楼）（侧楼）（面楼）

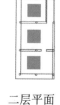

负一层平面（面楼）　二层平面（主楼）

以罗良海旧宅为标准模板，共分出了10个主功能模块，5个辅助功能模块。

- 辅功能模块
- 主功能模块

模块可能性分析

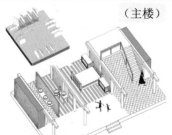

传统用做前厅和卧室

作为衣食住行生活中必不可少的住，属于主要的功能，有自住用的前厅或是做民宿的前台可供选择。

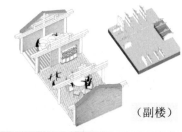

传统用做生产间或杂物室，可变为灶台+餐厅模块，等等

传统的大灶台还是现代化的厨房，餐厅是当地的小低板凳配矮桌还是游客习惯的餐桌，都是可以选择的。

成为其他主要功能可能性分析（分别列举三种可能性）

I 前厅/前台+卧室+厕所的可能性……

II 灶房/厨房+餐厅的可能性……

III 生产间+杂物间的可能性……

IV 活动间+火塘间

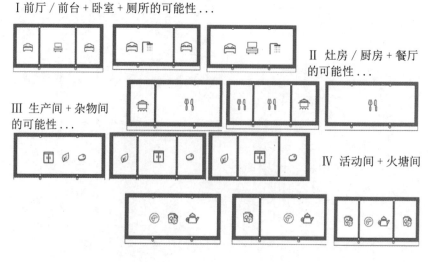

模块可能性分析

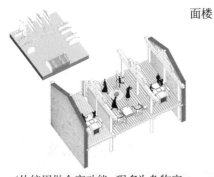

面楼

（传统用做会客功能，现多为杂物室，可变为火塘＋活动室模块），等等

火塘在当地人的地位是非常高的，同时当地农村人非常喜欢聚在一起闲聊，因此安排一个活动室就非常合适，可以同时满足游客和当地居民的双重需求。

成为其他主要功能可能性分析（分别列举三种可能性）

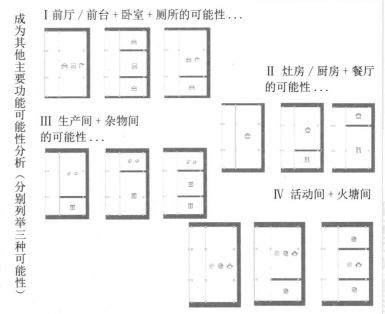

Ⅰ 前厅／前台＋卧室＋厕所的可能性...

Ⅱ 灶房／厨房＋餐厅的可能性...

Ⅲ 生产间＋杂物间的可能性...

Ⅳ 活动间＋火塘间

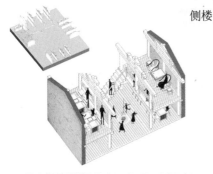

侧楼

（传统用做灶房和餐厅，现可用作餐厅或生产间）

当地人常用来制作腊肉，晾晒茶叶，或者其他农副产品的地方，必不可少。

成为其他主要功能可能性分析（分别列举三种可能性）

Ⅲ 生产间＋杂物间的可能性...

Ⅳ 活动间＋火塘间的可能性...

多项辅助功能

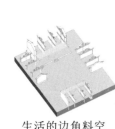   

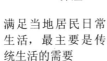

| | 养殖 | 小展厅 | 公共洗衣房 | 超市 |

生活的边角料空间的植入 / 满足当地居民日常生活，最主要是传统生活的需要 / 可能举行一些乡村文化、乡村摄影展、乡村农副产品展览等 / 满足长期住宿旅客的需求 / 满足城市居民日常所需商品需求，增加便利性

组合原则 　人数＋改造目标（自主或经营）＋动静态空间选择＋功能匹配度

项目落地所需技术手段

 保温隔热　　防虫防腐　　防噪音　 防雷击　 消防　 卫生条件　　防震

哈尔滨工业大学
Harbin Institute of Technology

文化寄居——哈尔滨道里文化宫空间改造设计
Cultural Sojourn

小组成员：陈博文 韩秋 张宇淳 朱雪莹

概念生成 | Concept Generation

开发文化价值

社区服务职责

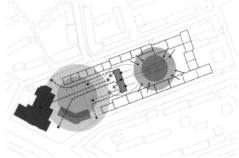

场地策略 | Site Tactics

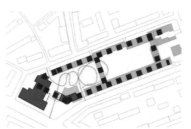

1. 打造连续的室外动线及步行系统，还原街区原本肌理。

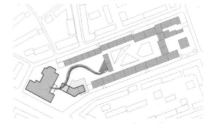

2. 用二层连廊串联并整合街区内各栋文化建筑，形成驻留空间。

3. 设置地上、下沉两个广场，形成驻留空间。

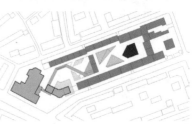

4. 将老年活动等社区功能进行整合，主要设置于地下空间，避免影响建筑采光。

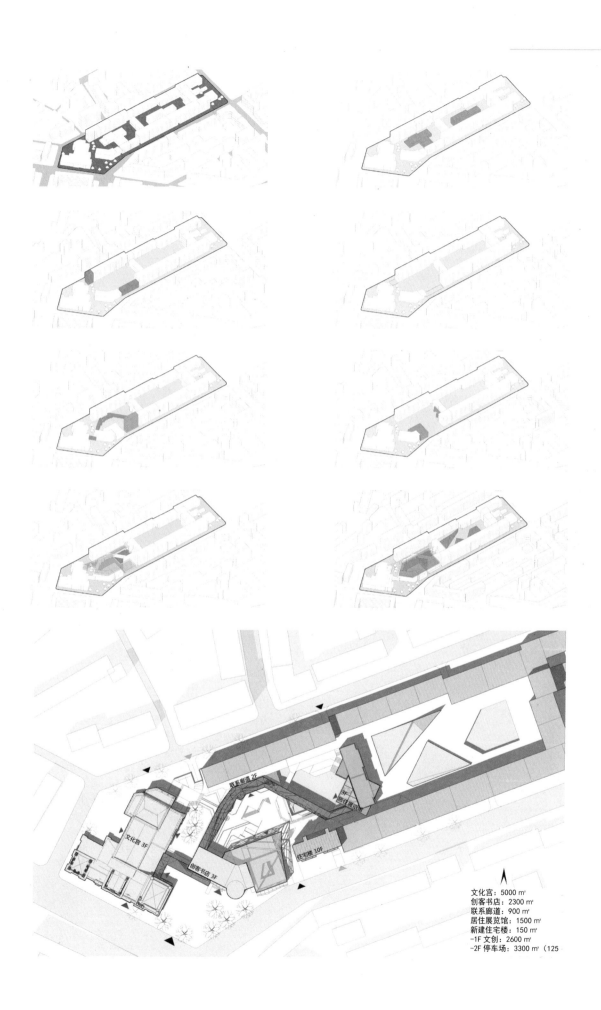

國粹賡續卷——歷史建築保護與再利用設計

中期檢查

文化宫：5000 ㎡
创客书店：2300 ㎡
联系廊道：900 ㎡
居住展览馆：1500 ㎡
新建住宅楼：150 ㎡
-1F 文创：2600 ㎡
-2F 停车场：3300 ㎡（125

单体设计 | Monomer Design

文化宫 | Cultural Palace

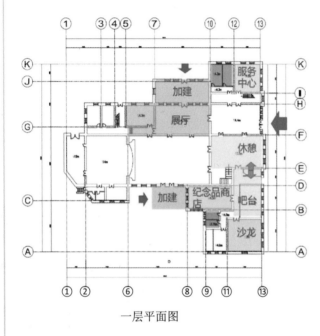
一层平面图

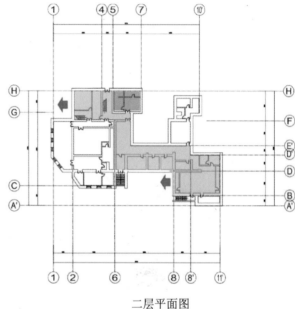
二层平面图

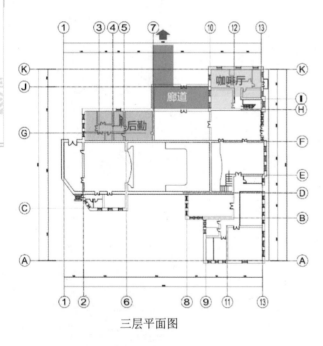
三层平面图

由于文化宫的文物保护等级很高，内部的很多格局和装饰，设计都尽可能地予以保留，并尽量采用可逆式的装置介入平面、剖面，完成功能改造和更新。

为了与户外环境产生更多的联系，我们重新设置了一层入口的位置，使环境和内部厅堂形成有机的整体，也让游客的流线更为顺畅。我们将旧有的办公区域改造成面向大众开放的餐饮、沙龙、商店等业态，目的是增强文化宫的商业性和公共性，提高文化宫的人气。二层局部打通了墙面与联系廊道相连，形成大空间，并和东侧的老年活动中心室内空间相互通达，形成功能和空间形态上的有机整体。文化宫的地下空间也与街区的地下相连，同地上二层的廊道、地面的广场一同形成立体的交通和功能网络。

创客书店 | Maker Bookstore

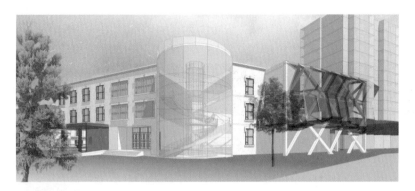

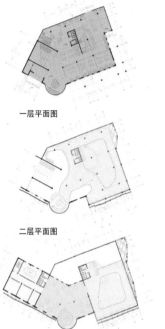

一层平面图

二层平面图

三层平面图

联系廊道 | Connection Hallway

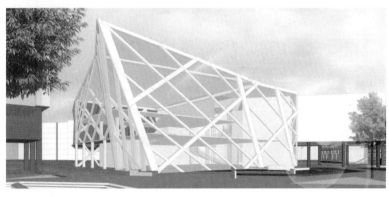

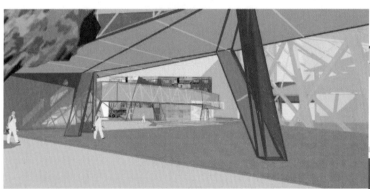

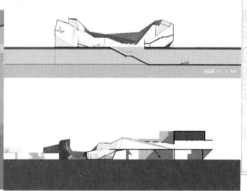

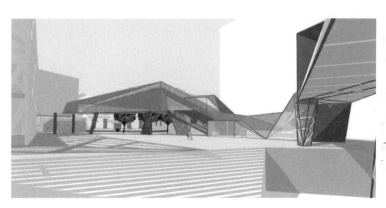

创客书店 | Maker Bookstore

居住博物馆 | Habitable Museum

纵墙承重

增加大空间

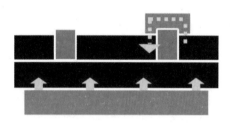
解决流线

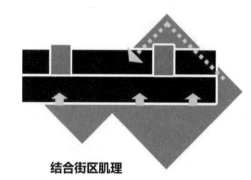
结合街区肌理

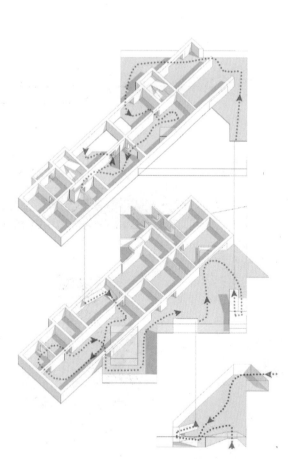

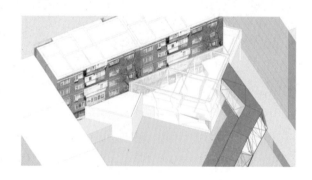

地下公共空间 | Underground Public Space

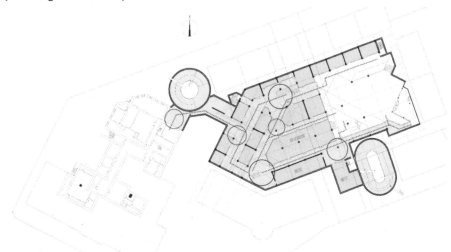

地下停车场 | Underground Parking

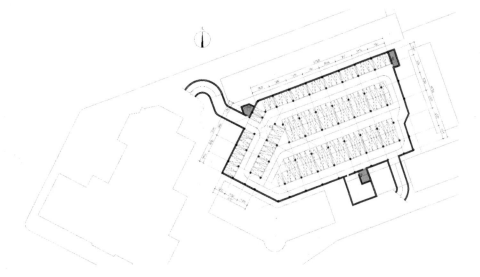

景观设计 | Landscape Design

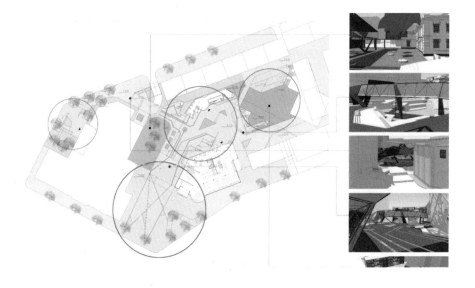

哈尔滨工业大学
Harbin Institute of Technology

历史建筑保护背景下建筑空间及其社区环境改造设计
Renovation Design of an Architectural Space and its Surroundings in the Context of Historic Building Conservation

「室内设计 6+」2018（第六届）联合毕业设计
"Interior Design 6+"2018(Sixth Year) Joint Graduation Project Event

小组成员：周子钦　颜岩　赵斌　周毓

基地概况 | Base Overview

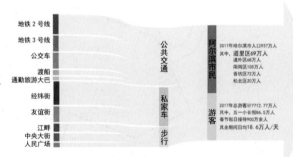

2号线、3号线地铁载客量：参考地铁1号线：16万人/天
渡船每日载客量：1.1万人/天
哈尔滨市公交车客流量：249万人/天

本方案把道里区文化宫作为市中心的一个标志性的剧场空间加以设计。方案研究和探索了一座传统建筑是否可以复苏且发挥更大潜力的命题。

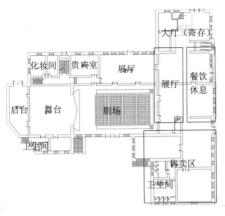
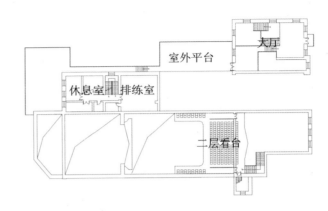
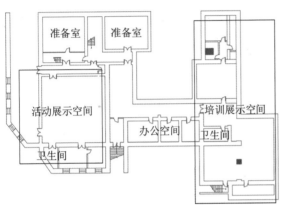

设计意图将文化宫周围的人民广场、中央大街、规划院、友谊宫、防洪纪念塔和江畔景观聚拢起来，把老道里区核心区域打造成文化及音乐的密集中心。这个新的区域将成为曲艺剧场文化场所。

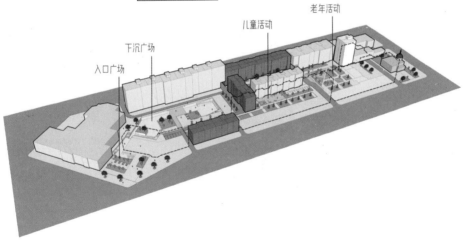

设计策略 | Design strategy

分析与改造 | Analysis and Renovation

梁氏古堡共有四个建筑群——御史院、大夫第、永宁堡、知府院。每个建筑群都有着各自的特点,我们根据每个建筑在所处地理位置的不同及各自内部空间特征的差异,并且考虑当地人群的生活需求,重新设计了梁氏古堡的建筑功能。

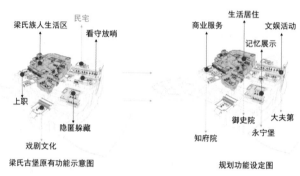

御史院 | Censor Courtyard

御史院共有九个院落和一个祭祖堂,是原先梁氏族人的聚集地,简称"九院一堂",御史院当下有六户在居,因后来买卖将御史院空间进行了分割,破坏了原有的院落格局,为此我们对御史院空间进行复原,拆除原有的分割墙体(现已无作用)。

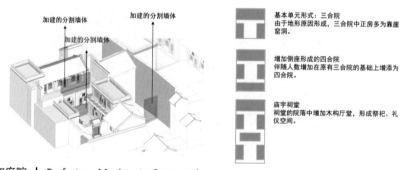

知府院 | Prefecture Magistrate Courtyard

知府院位于村落中心地点,周边道路交通顺畅,内部空间划分明显,格局恢弘大气。为此我们根据其所处位置及村落功能设置,我们对知府院的定位是餐饮服务功能。

大夫第 | Manor Garden

大夫第为原有的住守防御、放哨坚持、锻造囤放兵器之所在"上四下三",百尺巍楼便居于此,现已无人居住,原有的铸造设施及兵刃也已经不在,建筑结构破坏严重,已有两个院已经坍塌,原有的"堡中堡"建筑格局也已不再明显,我们根据其原有的场地功能,设置大夫第为文化活动区,在修复方面进行分类处理,根据不同需要,选择不一样的修复方式,只求重新激活场所精神、让文脉得以延续。

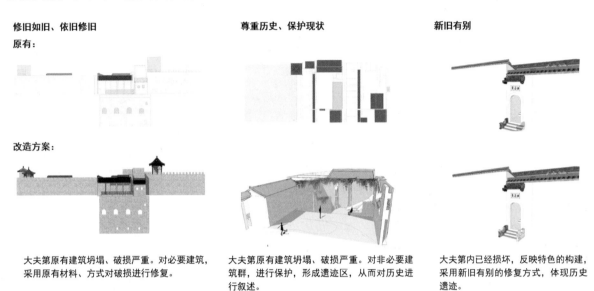

| 修旧如旧、依旧修旧 | 尊重历史、保护现状 | 新旧有别 |

大夫第原有建筑坍塌、破损严重。对必要建筑,采用原有材料、方式对破损进行修复。

大夫第原有建筑坍塌、破损严重。对非必要建筑群,进行保护,形成遗迹区,从而对历史进行叙述。

大夫第内已经损坏,反映特色的构建,采用新旧有别的修复方式,体现历史遗迹。

永宁堡 | Yongning Castle

永宁堡是原有重要的防御性建筑,与其他防御性建筑不同的是,永宁堡以防御功能为主导,其内部布满"黑、窄、深、陡"的暗道,根据永宁堡独有的建筑空间,我们对此的功能定位是记忆馆,用以记录当地原有的文化历史,供人参观,使文脉得以传承,文化得以延续,历史得以铭记。

庭院绿化　　　　样板展示间　　　　展台位置　　　　展示内容

鸟瞰图 | Bird's-eye View

西安建筑科技大学
Xi'an University of Architecture and Technology

韩城古城的保护与再利用设计
Conservation and Reuse Design of Hancheng Ancient City

「室内设计6+」2018（第六届）联合毕业设计
"Interior Design 6+"2018(Sixth Year) Joint Graduation Project Event

092

小组成员
曹玥玲
王玲子
任晓贤
张景一

场地分析 | Site Analysis

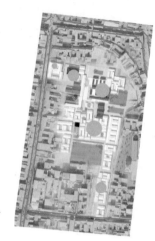

区域划分　　　　　　　定点

激活　　　　　　　　　古建

人群需求 | People's Demand

修复策略 | Renovation Strategies

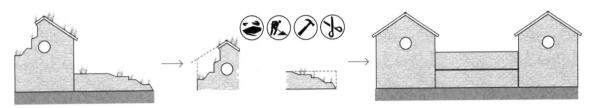

1. 部分受损与废弃房屋

2. 倒塌房屋与围墙

街道灰空间 | Gray Street Space

还原原始街巷结构

 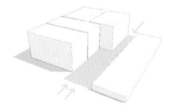 

街巷绿化

 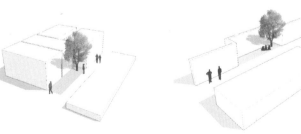

北京建筑大学
Beijing University of Civil Engineering and Architecture

恭王府博物馆展览设计·晋作家具制作技艺精品展
Exhibition Design for Prince Kung's Museum—Exhibition on Shanxi-style Furniture Craftsmanship and Masterpieces

「室内设计 6+」2018（第六届）联合毕业设计
"Interior Design 6+"2018(Sixth Year) Joint Graduation Project Event

小组成员：刘宇麒　闫铮　曹志玮

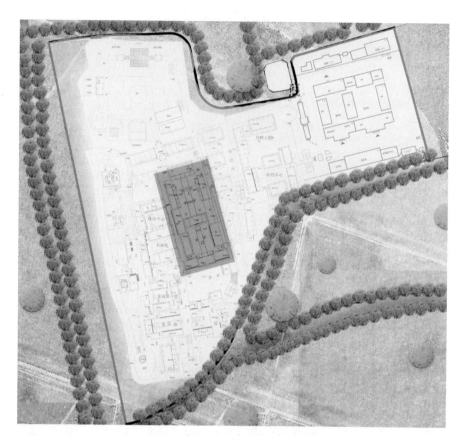

"北锁钥"——祖传汉高祖从大同领兵突围至此，六军欣然如归，"欣"同"忻"，"木欣欣以向荣，泉涓涓而始流"，忻州之名由此而生。

匠门溯源是故事的开始，我们将其安排在白特门。

通过时间线梳理晋作家具的闪光点，承接历史，有来源有去处，所以我们将匠心独运展厅放在百特门与桂香殿中间的仪鸾殿，有承上启下之意。

工序工具木材的总结梳理讲述的是家具背后制作的故事，有着至关重要的作用，桂香殿内部是一个整体，没有分隔的空间，我们将匠人材器安排于此。

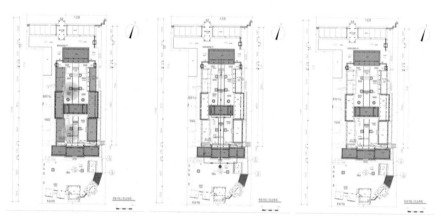

建筑分布图　　　流线分析图　　　展厅分布图

时期	秦汉	魏晋南北朝	宋元	明	清	清末民国	现代
发展	起源		趋于成熟	趋于成熟		延续、没落	发展保护
历史背景	秦汉时期在山西设立郡县，农业处于全国领先水平，促进了农具的发展，手工业的发展水平也是空前的	这一时期社会大动荡，大分裂，因其特殊的地理原因，呈现出军阀混战、狐族政权、频繁更替的交织状态，阻碍了这一时期的发展	北宋完成了汉唐主要地区的统一，山西再一次至于统一王朝的统治之下，政治建设、经济发展、社会局势为山西的发展创造前提条件	明清时期我国古代历史的发展达到顶峰，无论是表象的历史发展还是深层的制度演进，乃至人们的思想意识形态，整体的文化气质，都有集古代之大成的迹象。明初的古槐移民、叱咤商界的晋商、名扬后世的文化学人、蜚声海外的古城与建筑，都对明清的发展起到了重大的作用。			明清时期我国古代历史的发展达到顶峰，无论是表象的历史发展还是深层的制度演进，乃至人们的思想意识形态，整体的文化气质，都有集古代之大成的迹象。明初的古槐移民、叱咤商界的晋商、名扬后世的文化学人、蜚声海外的古城与建筑，都对明清的发展起到了重大的作用。
代表作品	陶寺晋作家具			（明）黄花梨六柱式架子床	（清）紫檀木象鼻牙头大画案		
传承方式			家庭承袭、秘籍单传、口传身受	家庭承袭、秘籍单传、口传身受	家庭承袭、秘籍单传、口传身受	家庭承袭、口传身受、师带弟承	
风格特征				（明式）造型厚重。虽形制上仍采用"明式"，但雕饰开始繁缛富丽	造型笨重、文饰呆板、梆框平、少起线		现代古典家具（新派晋式古典家具）
影响因素	历史文化		历史文化	历史文化、社会稳定情况、"主家"喜好、地域差异	匠人工艺水平稂莠不齐、地域差异		

研究晋作家具，由曹运建入手，曹运建是第四批国家级非物质文化遗产项目代表性传承人，他在20世纪80年代帮人收购古典家具时，对"晋作家具"产生了浓厚兴趣。对曹运建所带领的唐人居进行谱系梳理，我们可以挖掘出明清时期的木匠、油漆匠以及雕刻技师的创作特点，也能看出随着时代的变迁，晋作家具传统技艺的传承与发展。我们将人物谱系、故事背景与时代背景相结合。

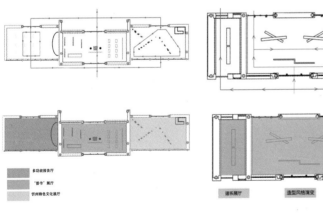
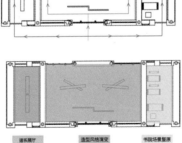
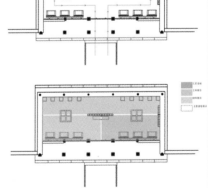

对传承人脉络谱系梳理，我们可以挖掘出明清时期的木匠、油漆匠以及雕刻技师的创作特点。沿波逐源，我们将人物谱系，故事背景与时代背景相结合，从而使体验者更好的走进晋作家具。这便是第一展厅的由来"匠门溯源"。

通过匠门溯源拉出的时间线进行谱系梳理，历代工匠独具创新、心思巧妙，将各自所处时代的历史特色都融入了自己的创作中，研究对比家具造型风格演变，提取出传承发展史中的闪光点，布置为另一展厅"匠心独运"。

"材"与"器"都是家具制作技艺中必不可少的基础，选材下料、设计画线、锯刨凿刻、油漆抛光各道工序都各有讲究。我们将工序工具木材都进行了总结梳理，这有了我们的第三展厅"匠人材器"。

北京建筑大学
Beijing University of Civil Engineering and Architecture

恭王府博物馆展览设计·广式家具制作技艺精品展
Prince Kung's Palace Museum—Exhibition Design for Guangdong-style Furniture Craftsmanship and Masterpieces

「室内设计 6+」2018（第六届）联合毕业设计
"Interior Design 6+"2018(Sixth Year) Joint Graduation Project Event

小组成员　韩玥　杨玉萍　郑一霖

项目定位 | Project Positioning

本次毕业设计课题围绕传统家具制作技艺国家非物质文化遗产保护与传承项目需求，踏勘北京恭王府，梳理广作家具传统制作技艺谱系，遴选代表性谱系分支，策划广作家具专题展览。总体设计原则将历史建筑的保护与再利用统一起来，处理好空间关系，做到"展院合一"，开展文物建筑保护与展陈利用设计实践。

基于恭王府的保护规划，探讨设计范围内传统古建筑的保护与利用方式以及更新设计方式，创造新式的传统技艺传承的展陈设计方式。

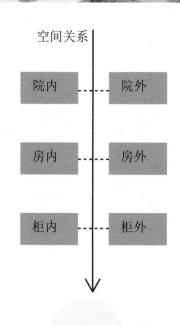

展览空间规划 | Display Space Planning

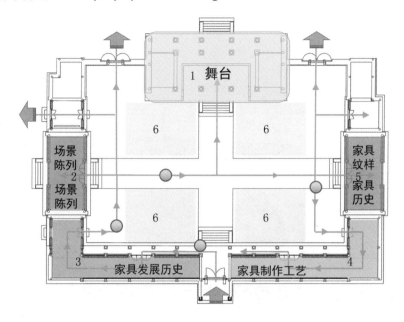

1. 东二府门
2. 西厢房
3. 西倒座房
4. 东倒座房
5. 东厢房
6. 草地

展陈大纲 | Display Outline

- **历史迷踪**

  从15世纪到18世纪，世界市场扩大，广州由于地理位置便利成为了中国与欧洲重要的贸易往来之地，受到当时政府高度重视。文化也同样受到欧洲文化的影响。当时广州人大多以出口家具谋生，其中就包括杨氏家族（介绍家族谱系、传承人口述史、陈列手稿、设计场景还原）。

- **纹样追溯**

  中国古典家具纹样历经几千年的发展和演变，形成了独特的艺术风格。商周时期的纹样神秘威严，春秋时期的浪漫绚丽，秦汉时期的飞动精炼，而清代的纹样繁缛富丽。这一风格脉络，传达表述着中华民族深厚的文化底蕴。（器型上演示纹样）

- **纹样追溯**

  无论是花卉草虫，还是人物故事，都给今人留下了古代丰富的生活气息和人文含义。即使是那些抽象的行云流水和几何图案所表达的，也是民族长期以来积淀的文化传统和精神。可以说它们作为文化的承载者，讲述着区域文化的差异性。

- **工艺工序**

  广式家具从清代中叶形成自己的风格起，便一直是"清式"家具的典型款式之一。追求用料充裕，大面积雕刻及镶嵌艺术。构件一般不拼接，而是一木制成，讲求木性一致。另外，装饰花纹雕刻深湛，刀法娴熟，磨工精细，纹饰表面晶莹如玉，不露刀痕。

南京艺术学院
Nanjing University of the Arts

万花筒——民国时期人物展
Kaleidoscope——Personages of the Republican Period

「室内设计 6+」2018（第六届）联合毕业设计
"Interior Design 6+"2018(Sixth Year) Joint Graduation Project Event

小组成员：
马悠庭
范聪达
高金宇
孙　昱

项目定位 | Project Positioning

　　晚清开始，民国时期思想受西方文化的冲击与中西文化的交锋，出现了一个高潮，产生了一大批思想家和众多大师级的人物文学家艺术家，可以说是继诸子百家在春秋战国时期后的又一次百家争鸣。民国，就是一个这样光怪陆离的魔幻大时代。民国人物就如同万花筒，因此我们借用万花筒的来表现民国时期的人物。我们做的就是关于民国时期传奇人物展览。

基地调研 | Base Research

场地选址

人流走向

周边建筑

交通概况

现有建筑调研 | Existing Building Research

原先的展览空间太过于死板，8个建筑全部选用照片展览的方式，过于单一。陈设摆放方式过于正规，丧失了参观的乐趣，缺乏活力。出口过多，容易让参观者迷失方向。

展示大纲 | Display Outline

	展厅名称	展示方向	展示内容	具体人物
第一展厅	一抹背景，悠悠水长	政治	民国总统，民国总理	孙中山
第二展厅	明月独举，气节长存	军事	民国将领，科学家	张学良
第三展厅	背影渐远犹低徊	艺术	民国电影，戏曲，民国建筑师	梅兰芳
第四展厅	真名士自风流	文化	著名文人	辜鸿铭

立面图 | Elevation Plan

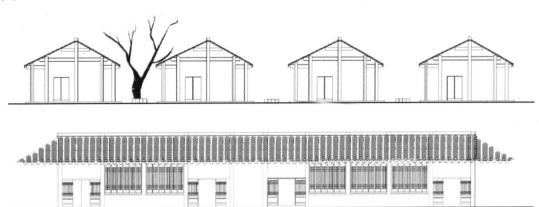

浙江工业大学
Zhejiang University of Technology

浙江省薛下庄村『二十四间』老建筑保护与再利用设计
Conservation and Reuse Design of "Twenty-four" Old Buildings in Xuexia Village, Zhejiang Province

「室内设计 6+」2018（第六届）联合毕业设计
"Interior Design 6+"2018(Sixth Year) Joint Graduation Project Event

小组成员：章家骐　章佳祺　刘叶　蒋一德

院落空间分析 | Courtyard Space Analysis

院落现状 | Courtyard Status

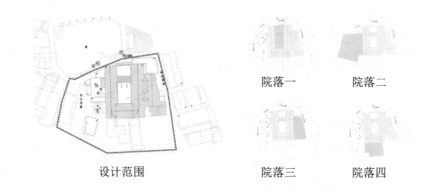

设计范围　院落一　院落二　院落三　院落四

改造策略 | Renovation Strategies

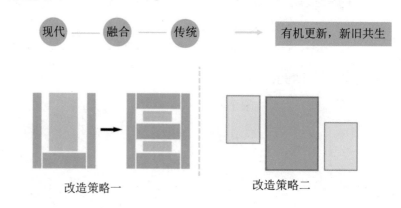

现代—融合—传统　→　有机更新，新旧共生

改造策略一　改造策略二

改造分析 | Renovation Analysis

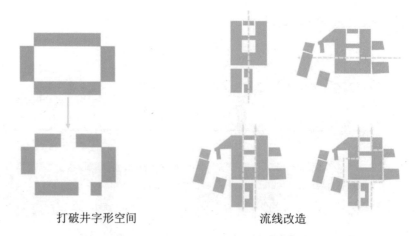

打破井字形空间　流线改造

总平面 | Master Plan

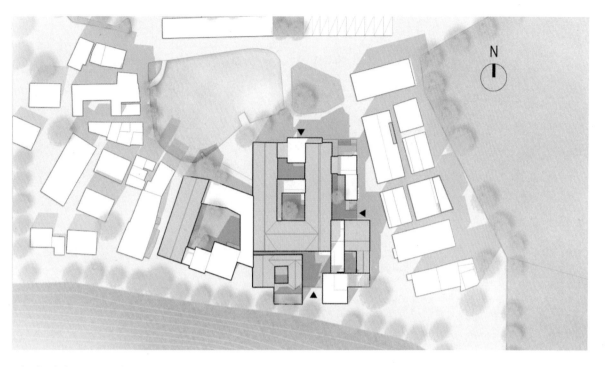

方案分析 | Scheme Analysis

庭院分析 | Courtyard Analysis　　　　　　　　　　　　流线分析 | Flow Line Analysis

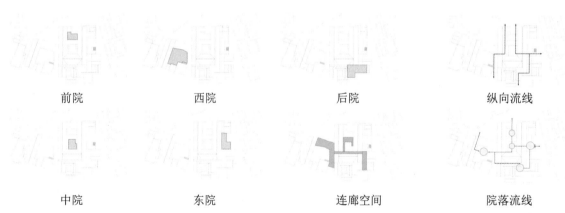

| 前院 | 西院 | 后院 | 纵向流线 |
| 中院 | 东院 | 连廊空间 | 院落流线 |

鸟瞰图

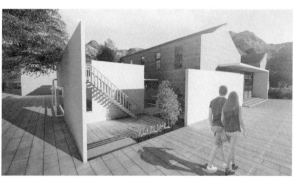

效果图

热点命题，尽显特色

联合指导，服务需求

答辩展示

"室内设计6+"2018(第六届)
联合毕业设计
"Interior Design 6+" 2018 (Sixth Year)
University Joint-enterprise Cooperative Graduation Project Event

基于城市更新背景下的综合体设计问题
海淀镇温泉中心城区功能提升设计研究课题

冬奥背景

同济大学图书馆室内外环境更新改造设计

Conservation and Regeneration Design of the Indoor and Outdoor Environment of Tongji University Library

高　　校：同济大学
College: Tongji Unversity

学　　生：李一丹、张迪凡、许可、马一茗、刘雨婷
Students: LI Yidan, Zhang Difan, XU Ke, Ma Yiming, Liu Yutin

指导教师：左琰、林怡
Instructors: Zuo Yan, Lin Yi

课题评价：优秀
Achievement: Excellent

学生感悟
Student's Thought

许可

　　建筑和室内设计不只是解决空间上的问题，我们通过一个学期的调查研究，与老师和各个实践建筑师的交流，以及与图书馆的管理方和使用者的交谈，我们逐渐发现，建筑和室内设计更应该关注的是人们的需求、建筑的使用方式和人们在其中的生活。此外，也要特别感谢指导老师一学期的耐心指导教授，以及组员之间的互相鼓励支持！

张迪凡

　　作为大学本科最后一场课程作业，本次课题从设计伦理层面的理念探讨到室内空间的节点构造全方位多层面均有涉及。让我们融会贯通本科以来的所学所思，从设计到建造的全过程，有了一次完整的体验，并最终和小伙伴们一起，在欢笑参伴也日夜不休中顺利完成了项目课题。最后诚挚感谢老师耐心细致的指导和队友齐心合力的并肩战斗！

李一丹

　　通过一个学期的学习，我对室内设计有了更深入的了解，并从更加专业的角度全面认识了学校的图书馆的重要性、建造过程以及未来改造方向，等等。感谢老师的指导和督促，也感谢小组成员的通力合作，希望未来还能够一起学习，共同进步。

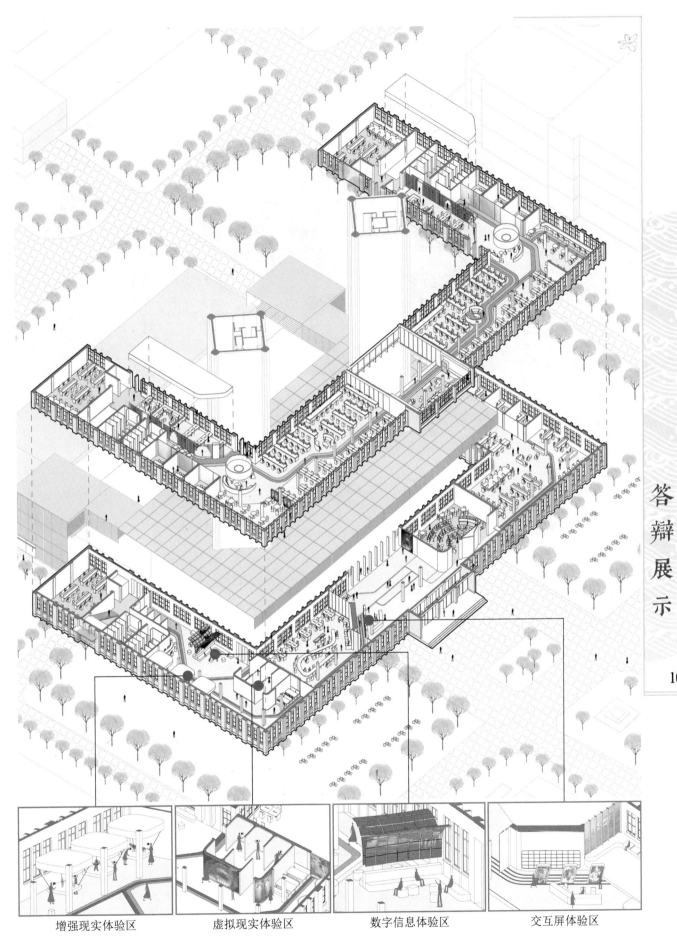

增强现实体验区　　虚拟现实体验区　　数字信息体验区　　交互屏体验区

國粹賡續卷——歷史建築保護與再利用設計

答辯展示

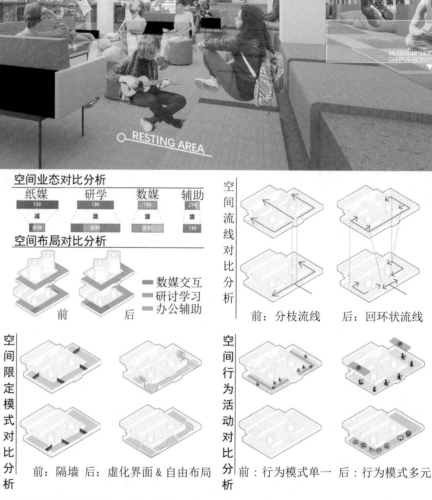

纸媒至上的图书馆模式在当今时代日益衰退，数媒交互体验的高效开放与便捷使得知识传播的途径趋势开始发生变化。该设计旨在探索当代图书馆"以人为中心"的空间模式。

U形群房一层作为流通量极大的校园风采视窗，通过顶面立面等各个空间界面进行数字媒体的空间传达，并构筑多模式的空间使用场景，以促进人与信息界面／人／空间的交互，加强校园文化的传播与凝聚。

二层作为典型的学习空间，探讨了一种新的弱化纸媒空间占有的"阅览室模式"。将书籍封装，吊于屋顶，采用自动售卖机的借阅模式，在终端输入书籍信息，系统便会为你将其调出。最大化提升纸媒检索效率的同时充分解放了下部人的活动空间，达成一种全新的室内空间模式。

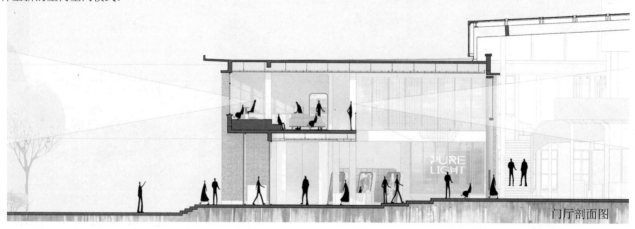

门厅剖面图

一层 信息开放窗口

- 文化展示传播**薄弱现状**

校园文化活动等展示宣传界面	无
交互体验式的学术浸润空间	无
面向外界公众的图书馆信息开放	无
数媒界面的学习体验空间	少

- 校园开放窗口空间打造

门禁
对公开放的学习

- 空间中的信息传递途径：
  - 虚拟现实
  - 增强现实
  - 全息投影
  - 交互屏
  - 计算机
  - 纸媒书籍

空间多模式 A

开放式学习
- 自由组合
- 交互共享
- 动态空间

临时式展览
- 传播校园文化
- 丰富信息模式

空间多模式 B

报告厅

开放研讨

空间多模式 C

普通办公
- 插板分隔
- 空间高效
- 私密性

自由办公
- 插板组合
- 讨论合作
- 空间灵活

知识媒体中心　开放学习/展览空间　办公区

一层平面图

國粹賡續卷——歷史建築保護與再利用設計

答辯展示

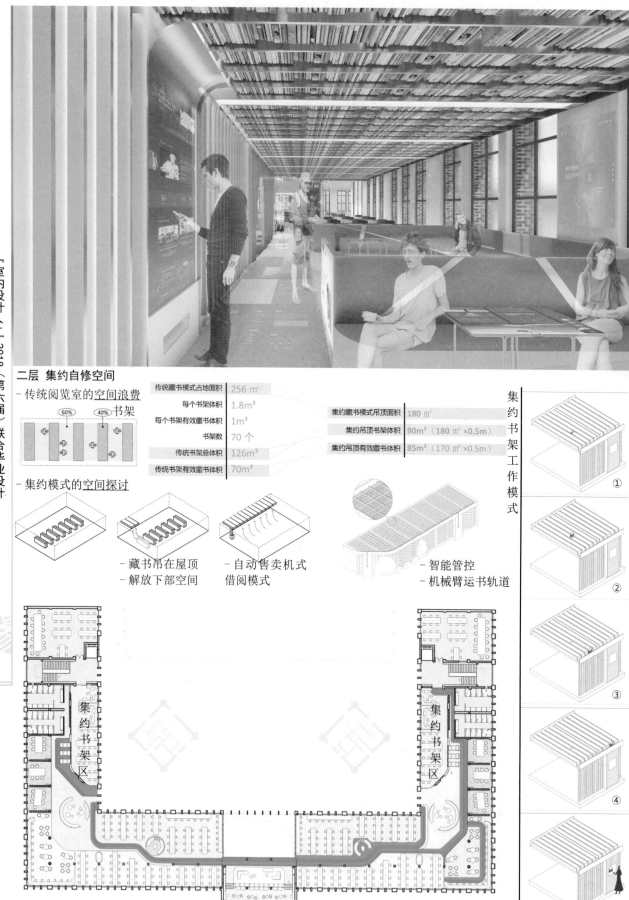

二层 集约自修空间

- 传统阅览室的空间浪费
- 集约模式的空间探讨

传统藏书模式占地面积	256 m²
每个书架体积	1.8 m³
每个书架有效藏书体积	1 m³
书架数	70 个
传统书架总体积	126 m³
传统书架有效藏书体积	70 m³

集约藏书模式吊顶面积	180 m²
集约吊顶书架体积	90 m³ (180 m² × 0.5m)
集约吊顶有效藏书体积	85 m³ (170 m² × 0.5m)

- 藏书吊在屋顶
- 解放下部空间
- 自动售卖机式借阅模式
- 智能管控
- 机械臂运书轨道

集约书架工作模式
① ② ③ ④ ⑤

二层平面图

集约书架区

大厅设计 | Lobby Design

多功能信息中心——图书馆大厅设计
| Multipurpose Information Center—Library Lobby Design

图书馆的大厅现为钢结构的椭圆形门厅，考虑到椭圆形门厅的一些问题和图书馆未来的发展趋势，我们拆除了椭圆形大厅，将裙房与书库围成的合院全部罩在了玻璃顶下，使之成为一个"庭院"，呼应历史原型的同时，最大限度保留历史立面，并将门厅扩大为一个完整的空间，在其中加入了前期调研中呼声最高的一些公共性较强的功能。

大厅功能变化

	借还书服务	展览	查询	交流	休闲	藏书	多媒体
改前							
改后							

改后大厅空间特点

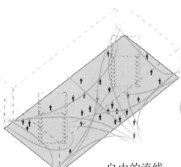

开敞的视觉关系

自由的流线

多样的行为

现实增强技术辅助设计只需一个简单的移动终端即可实现，包含位置共享、资源共享、辅助查找、情绪调节、借阅量查询等功能。

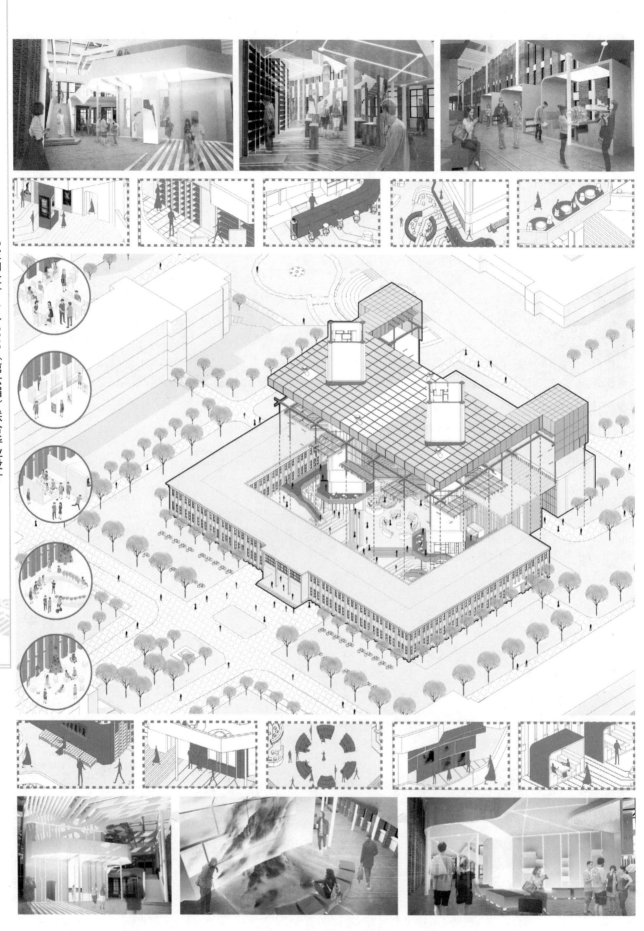

塔楼设计 | Tower Design

校园芯片——图书馆塔楼设计 | Campus Chip—Library Tower Design

对图书馆塔楼部分进行改造，首先需要对塔楼现存的问题进行分析。首先，从业态分布来看，塔楼区域与裙房整体较为割裂，五层办公空间将学习空间打断。核心筒内电梯狭小，导致空间垂直向可达性逐渐降低。功能分区单一，以自习区和阅览区为主，不能满足多种需求。从平面图来看，边角空间浪费严重，由此加剧了高峰期座位供不应求的现象。

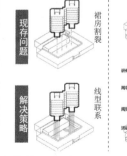
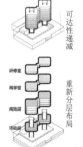

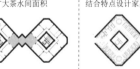

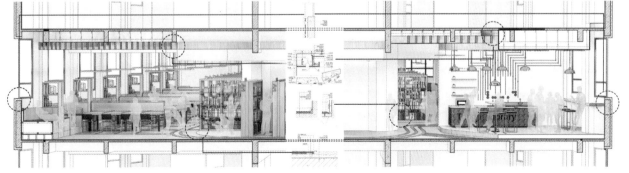

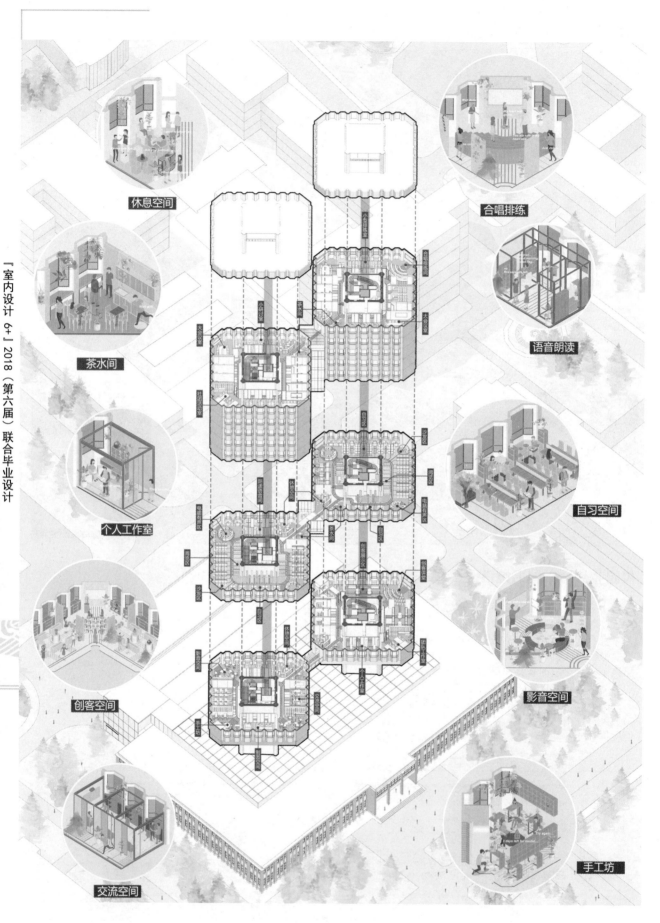

导师点评

周立军
Zhou Lijun

同济大学图书馆作为校园中轴线上一栋标志性建筑，承载着同济大学的历史变迁与发展。因此，以同济大学图书馆空间改造设计为题的毕业设计，是比较符合今年历史建筑保护与更新的课题方向的。设计方案在较深入的实地调研基础上，适应当今师生对图书馆功能的新要求，在传统图书馆空间模式基础上，增加阅览与信息交流的便捷性和空间的容量。以校园芯片为主题，诠释了空间改造的目标和理念。设计方案通过对图书馆裙房和塔楼中交流与阅览空间的梳理与整合，从问题出发进行改造设计，方案设计逻辑清晰，分析到位，图纸完成度高，特别是针对图书馆采光照明等关键问题，进行了软件模拟的定量研究，同时采用手工模型对重点空间推敲分析，这些都值得肯定。

Tongji University Library, as a landmark building on the central axis of the campus, carries the historical development of Tongji University. Therefore, the spatial transformation of Tongji University Library is a graduation project in line with this year's topic of historic building conservation and renewal. Based on in-depth field research, the designer adds a space of efficient reading and information exchange to the traditional library space to meet students and teachers' new requirements for library functions. The campus chip is taken as the theme, interpreting the objective and idea of space transformation. The designer makes the transformation design of the communication and reading space in the skirt building and tower building with the problems taken into consideration. The conceptual design contains a clear logic and proper analysis, and the drawing is elaborate. Especially, for the key problems such as daylighting and illumination in the library, quantitative research is done by software simulation, and meanwhile a manual model is used to analyze key spaces. All these are worth approving.

专家点评

宋微建
Song Weijian

项目方案完整的分析与展现了从"纸媒"到"多媒体"的进程演变以及这一变化对"阅读"产生的影响。

方案初期，针对"用途"和"用途的变化"两方面产生思考，进而采取了适用于现代生活方式的一系列安排：将图书馆中，单一的以"书"为中心的模式进行弱化；强化了一种全新的、高效、开放、便捷多元的阅读方式。通过这样探索性的功能覆盖以及强弱调整，不仅满足基本的阅读功能，也扩展出了"数媒""交互""体验"新的特性。

空间的把控与运用，是能够调动一切手段。不论是动线、布局，还是采光、照明，方案在空间的调配调动上，都为满足"新阅读模式"而服务。在空间设计上，表现突出的是数字芯电集成图，作为整个设计的大概念，它巧妙地把空间功能中繁复多变的不同要素，有机地整合在一起，使得空间既有序又有趣，同时富于灵动和变化。

The project completely analyzes and reveals the change of "print media" to "multi-media" and the effects of this change on "reading".

At the early stage of the project, the designer thinks about "usage" and "change of usage", and then make a series of arrangements applicable to the modern lifestyle: the single book-centered reading mode is weakened; a new efficient, open and multiplex reading pattern is promoted. Such exploratory functional coverage and adjustment of the strong and the weak not only generates basic reading functions, but also creates new features such as "digital media" "interaction" and "experience".

The control and application of spaces is a means by which everything can be mobilized. Whether it is a dynamic line, layout, daylighting or illumination, it serves the "new reading mode" through space deployment. In terms of space design, what is highlighted is a digital core electronic integration diagram. As a big concept in the design, it skillfully integrates different changeable elements in spatial functions together, making spaces not only ordered and interesting, but also flexible and varying.

华工附属中学旧建筑再生——高校众创空间设计
Regeneration of an Old Building in the Affiliated Middle School of South China University of Technology —Intramural Public Entrepreneurial Space Design

高　　校：华南理工大学
College:　South China University of technology
学　　生：汪瀚、邓康利、杨明之
Students:　Wang Han, Deng Kangli, Yang Mingzhi
指导教师：王琛、谢冠一
Instructors: Wang Chen, Xie Guanyi
课题评价：良好
Achievement: Good

学生感悟
Student's Thought

杨明之

参加这次"6+"比赛对我来说最重要的并不是得奖与否，而是在完成毕设的过程中与另两个组员之间结下的深厚感情以及指导老师对我们的谆谆教诲。

汪瀚

非常感谢这次比赛，在参与比赛的过程中，见到了国内高校的设计水平更是对自己的一种鞭策和刺激，让自己在以后的道路上会更加努力。

邓康利

设计是开卷考试，做设计时应该考虑更多的方面、更多的可能，不要局限在自己的小圈子里。

高校创新创业孵化基地 | Maker Space of University

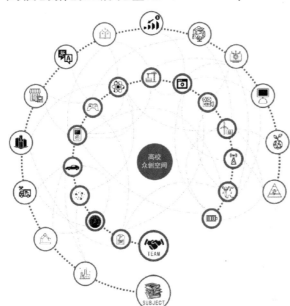

如左图所示，外圈为高校专业的图标，当多个专业人才汇聚在一起，便形成内圈的众团队，而众创空间便是一个提供让专业交错合作、让团队间相互合作交流的平台。

如下图所示，无人机、VR、生物感应等都是已有的先进技术，但它们碰撞在一起，可能会创造出新的技术类型——用VR操控的无人机技术。所以说创业团队间的合作交流，往往是迸发灵感、合作创新的催化剂。

团队合作示意图 | Teamwork Diagram

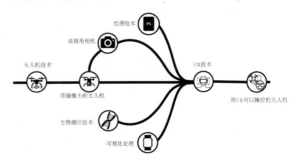

众创空间调研分析 | Public Entrepreneurial Space Research and Analysis

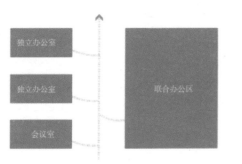

华工众创空间空间形态

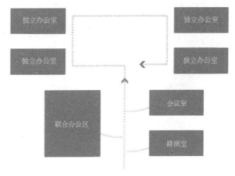

广大众创空间空间形态

我国高校众创空间大部分仍采用传统的办公模式，以办公功能区与辅助办公的会议、路演等功能为主，功能较为单一，缺乏公共空间，团队与团队之间几乎没有交流。

工作模式分析 | Working Mode Analysis

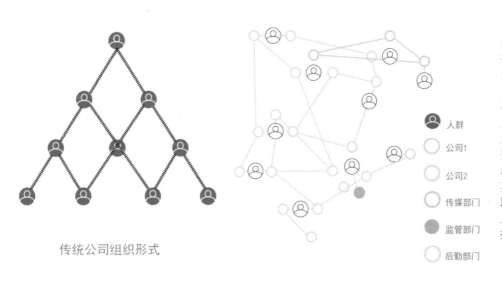

传统办公空间的组织以职能部门的直接连接而成，而众创空间由于一人多职和多人一职普遍现象的存在，所以职能与人员之间不是一对一关系，而是具有很强的交叉性和变化性。也就是说，创客公司在其人与其职务的直接联系下，同种职务将人与人的关系间接连接。

基地地理位置图 | Base Location Map

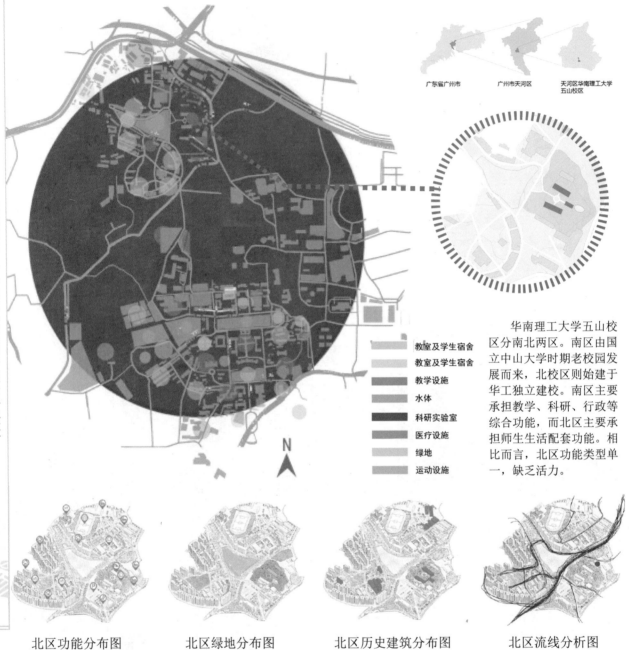

华南理工大学五山校区分南北两区。南区由国立中山大学时期老校园发展而来，北校区则始建于华工独立建校。南区主要承担教学、科研、行政等综合功能，而北区主要承担师生生活配套功能。相比而言，北区功能类型单一，缺乏活力。

北区功能分布图　　北区绿地分布图　　北区历史建筑分布图　　北区流线分析图

基地入口　　基地中庭
基地正视图　　基地一号楼　　基地二（三）号楼

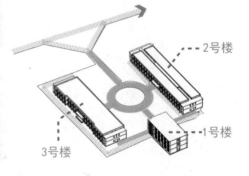

基地由三个独立建筑组成凹字形建筑群，流线比较单一，主要由圆形花坛组成。

众创空间定位与功能策划 | Public Entrepreneurial Space Positioning and Function Planning

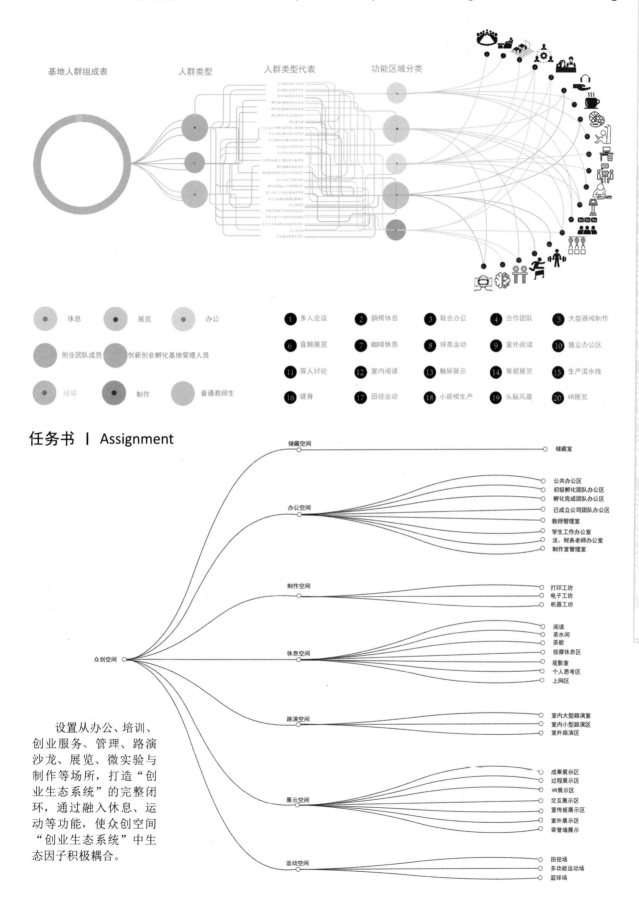

任务书 | Assignment

设置从办公、培训、创业服务、管理、路演沙龙、展览、微实验与制作等场所，打造"创业生态系统"的完整闭环，通过融入休息、运动等功能，使众创空间"创业生态系统"中生态因子积极耦合。

景观设计 | Landscape Design

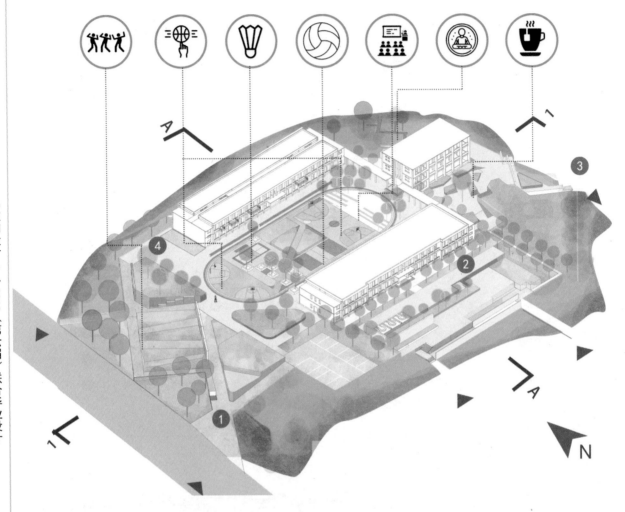

入口设计 | Entrance Design

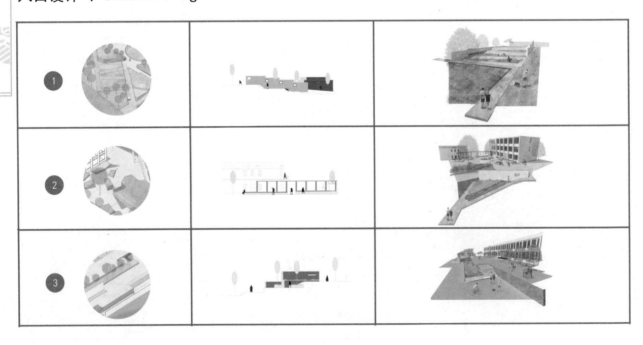

庭院设计 | Courtyard Design

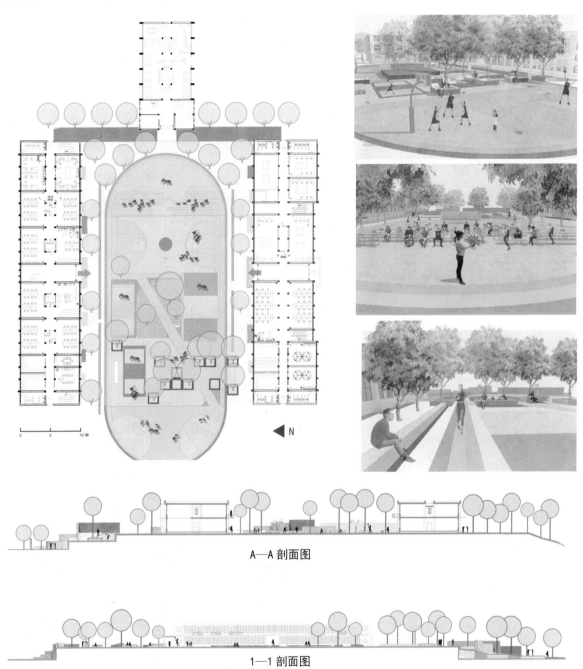

A—A 剖面图

1—1 剖面图

建筑室内外环境整体设计 | Overall Design of Indoor and Outdoor Environment

现有空间流线 | Existing Space Flow Line

室内功能分析 | Indoor Function Analysis

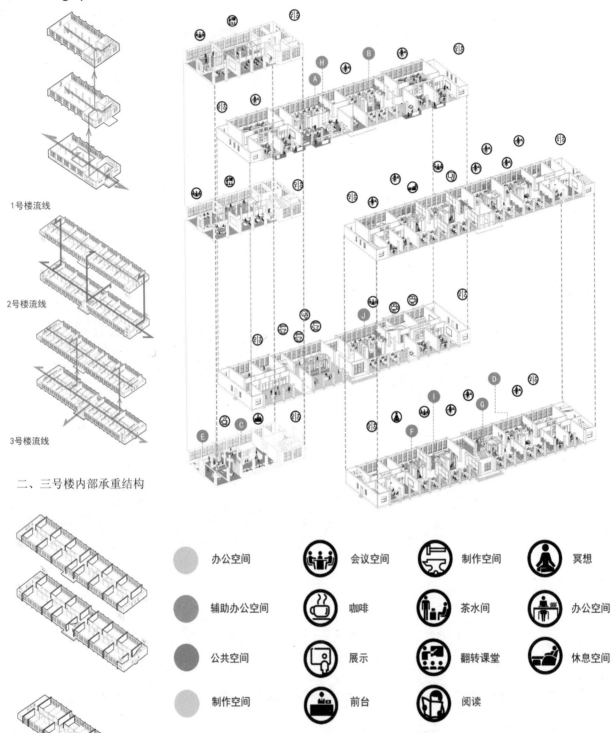

1号楼流线

2号楼流线

3号楼流线

二、三号楼内部承重结构

图例						
● 办公空间	会议空间	制作空间	冥想
● 辅助办公空间	咖啡	茶水间	办公空间
● 公共空间	展示	翻转课堂	休息空间
● 制作空间	前台	阅读	

　　为了贯彻前期调研中对众创空间现有问题的分析（现有众创空间仍采用办公模式，缺少交流），通过设置参与式运动场所（或设施）、共享混合的空间节点来尽可能地增加创客之间，创客与校园其他师生之间从偶遇到交流再到合作的可能，打造"创新共享社区"。

　　二、三号楼为砖混结构，建筑内部的砖墙是承重结构之一，不可改动。

节点分析 | Node Analysis

我们利用廊道的两排柱子,围合成一个个小盒子,从而打破鱼骨型空间布局,加强团队间在室内的交流。在办公空间里插入一些能让人停留的节点。

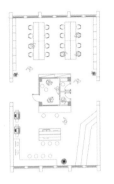
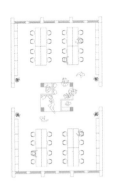
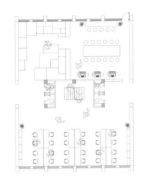
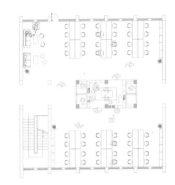

讨论空间

休息空间

阅读空间

复合空间分析 | Composite Space Analysis

让空间复合多种功能,从而用更优质的空间来吸引创业人员使用。

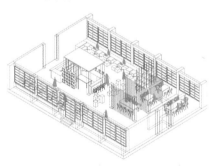
一号楼咖啡厅 休闲功能

一号楼咖啡厅 水吧功能

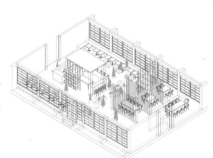
一号楼咖啡厅 展览功能

一号楼咖啡厅 书吧

一号楼咖啡厅 过道

一号楼咖啡厅 吧台

因为老建筑为砖混结构，我们或许改变不了它的物理空间形态，但我们可以通过设计来引导人的行为。二、三号楼作为办公楼，我们将打印、水吧等常用到的功能整合到复合空间中，且加入休闲书吧等功能，从而吸引创业人员使用，创造人与人之间、团队与团队之间交流的机会。

二号楼复合空间——水吧

二号楼复合空间效果图

三号楼复合空间书吧效果图

三号楼复合空间效果图

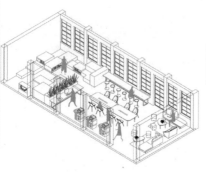

三号楼复合空间——打印

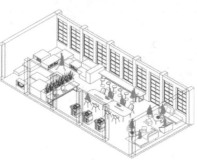

三号楼复合空间——自习

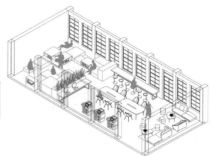

三号楼复合空间——书吧

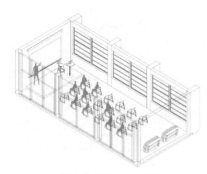

多功能教室——路演

多功能教室——翻转课堂

多功能教室——讲课

导师点评

杨琳
Yang Lin

华工毕业设计任务是将华工五山校区北区一组由三栋六十五年历史建筑围合的组团，改造成为校园众创空间，完成历史建筑保护再利用，解决由于南北校区功能和发展不均衡带来的尴尬，原有两层砖混结构空间划分单一的限制，吸引人流动线不便与建筑群环境优美静谧的矛盾等现存问题，通过赋予新的功能来改变现状、换发新生，来满足众创空间多种人员组合交叉性、多变形、多用途，以及分隔独立与协同融合的各种需求。

本组同学通过深入调研和切身使用感受，意欲突破北区为生活区的历史沿革，因势利导变成吸引学生创业交流的优势，运用运动跑道视觉元素，将室内外空间及景观和视觉与人结合，让建筑群的每个空间都变成看得见风景的空间。形成建筑空间环境的创业生态系统、辅助创新共享社区的生态系统构建，实现学生设计初心目标。

建筑本体因建筑结构的局限，对竖向空间的交通、动线、视觉的突破不够，对共享空间的多功能用途设计不够充分，应打开一部分楼板，创造集会展示等多功能空间，便于日常宣传路演交流活动。

South China University of Technology's design task is to transform a 65-year-old building complex in northern Wushan Campus of South China University of Technology into a collective entrepreneurship space, to achieve the purpose of protecting and reusing the historic building, to solve of the problem of the disparate development of the southern and northern parts. and other existing problems such as monotonous partitioning of the brick-concrete two-storey space, and limited pedestrian flow in the beautiful, tranquil building environment. After given novel features, the building complex takes a new look and can be used by different people as a public entrepreneurial space for multiple purposes, to meet various needs for spatial partitioning and collaborative work.

After intensive field research, the design team wants to transform this living area in the north zone into a space where students can exchange ideas with one another on collective entrepreneurship. So, by using the visual elements of running tracks, they combine the indoor and outdoor spaces and landscape with people's vision, making every space in the building complex a space where landscape can be seen. In this entrepreneurial ecosystem in the architectural spaces, an ecosystem of innovation and sharing can be built to help students make a design of what is in their heart.

The building has a limited architectural structure, which is bad for vertical spatial traffic, flow and sight, particularly for sharing. So, some floor slabs should be opened to form a multi-purpose space for makers to work together, or run promotions, hold road shows and carry out interchange activities.

专家点评

幸晔
Xing Ye

设计建立在大量的调研和分析的基础上，深度分析了众创空间的使用需求，较为准确地把握了使用人群的行为和心理，并把分析的结果运用在设计方案中，考虑了办公空间灵活的使用方式，抓住了运动这一主题，在室内以及环境景观方面都做了精心的设计，把几个六十多年历史的老建筑改造成一个轻松、活跃、适合年轻人工作、学习的办公空间，在保护历史建筑的同时提升了老建筑的使用价值，总体来讲，设计目标明确、逻辑清晰，但在设计深度方面稍显薄弱。

The design, based on an enormous amount of research and analyses, deeply analyzes the demand for a public entrepreneurial space, accurately grasps users' behavior and mentality, and applies analysis results to design. The designer considers the flexibility of use of office spaces, gets the main idea of sport, and makes an elaborate design for the interior space and environmental landscape. Several old buildings built over 60 years ago are transformed into relaxing, dynamic spaces for young people to work and study in, improving the use value of the old buildings while protecting them. Overall, the design is targeted, and the logic is clear, but the depth of design is shallow somewhat.

文化寄居——哈尔滨道里文化宫空间改造设计
Cultural Sojourn —— Harbin Daoli Cultural Palace Spatial Reconstruction Design

高　　校：哈尔滨工业大学
College: Harbin Institute of Technology
学　　生：陈博文、朱雪莹、张宇淳、韩秋
Students: Chen Bowen, Zhu Xueying, Zhang Yuting, Han Qiu
指导教师：周立军、赵翚
Instructors: Zhou Lijun, Zhao Hui
课题评价：优秀
Achievement: Excellent

学生感悟
Student's Thought

陈博文

我从这次毕业设计中又认识到了许多需要继续学习的东西，在以后的道路中会更加努力前行。

朱雪莹

大学五年收获太多，学到了许许多多的专业知识，也磨砺了性格，结识了来自五湖四海的伙伴，每个人身上都有自己独特的闪光点。我对于建筑有了全新的认识，并对以后的道路产生了思考。

张宇淳

在调研过程中了解到要运用多种途径方法来更深入获取更贴近当地居民的生活信息，才能设计出真正具有在地性的建筑。

韩秋

感谢在哈尔滨工业大学的五年，让我们收获了友情、理解、青春、乐观和团结。在这个温馨的大家庭中，我成长了许多，感恩母校，感恩建筑学。

小组平面 | Team Plane

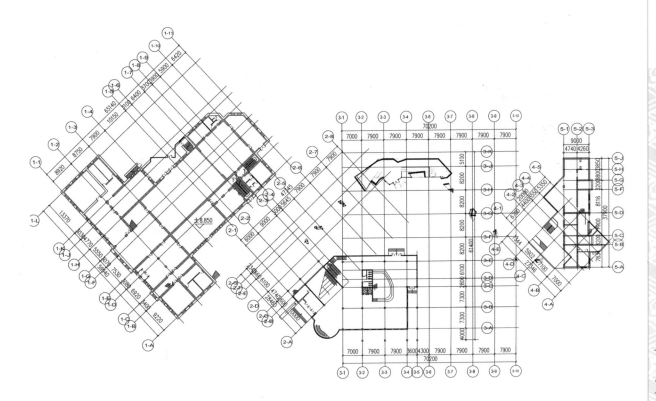

一层平面图

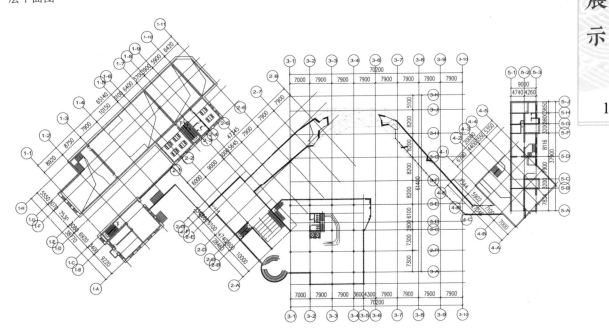

二层平面图

文化宫室内设计 | Interior Design of Cultural Palace

剧场观众厅 | Theatrical Auditorium

剧场观众厅室内效果图

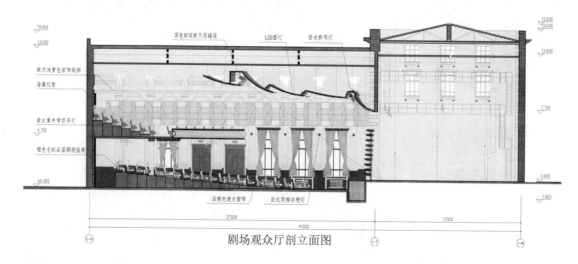

剧场观众厅剖立面图

咖啡厅 | Cafe

咖啡厅室内效果图

休息大厅室内设计 | Interior Design of Lobby Lounge

次厅从原本的办公空间改为休闲水吧，把原本分隔两厅的内墙打通，使其成为连接主厅和纪念品商店的消费休息区，增加可容纳的人流量。主要的装饰性家具是吧台柜，也是该空间的视觉焦点。折线型的背板富有动感，增强了空间主视点的纵深感，引人走近；用餐区上方倾斜的"满天星"木吊顶为该空间增加了优雅奇幻的光环境；拆除了空间尽头墙面的木墙裙，使整个墙面变得完整，再贴上文字使墙面设计简约而大气。

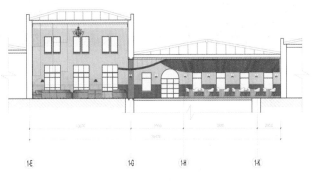

休息厅室内效果图　　　　　　　　休息厅室内剖立面图

展厅室内设计 | Interior Design of Exhibition Hall

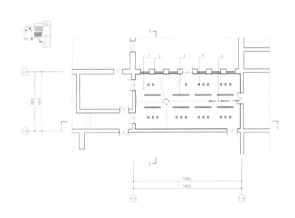

展厅室内效果图　　　　　　　　展厅室内家具布置图

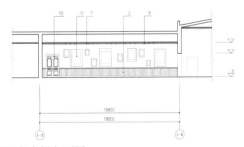

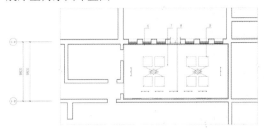

展厅室内剖立面图　　　　　　　　展厅室内灯具布置图

空中廊道室内设计 | Interior Design of Air Corridor

廊道形体的概念有二:一是以舞蹈表演为基点,如同一条红色的飘带轻盈地连接着各个建筑物;二是象征街区的脉搏,用简洁有力的折线形体表达新文化的活力与自信。从形态、材质、颜色、结构都和道里文化宫产生鲜明的对比,相互映衬。

动感的线条在立面上交织错动,有疏有密,形成丰富的节奏韵律。红色为场地中唯一的亮色。

涂鸦展廊作为新文化的展示平台和作为传统文化代表的文化宫进行碰撞。我们增加了一些和围护结构同样的三角形元素作为涂鸦板,使整个空间更简洁和谐,强调了结构线条感的同时,也使灯具作为装饰构件,和整个空间融为一体。

涂鸦展廊室内效果图

休息厅室内效果图

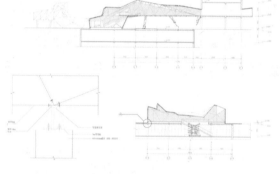

节点详图　　剖面图

居住博物馆室内设计 | Interior Design of Habitable Museum

居住博物馆二层室内效果图

红砖楼是一栋住宅楼,其原本的属性决定它内部空间十分狭小,交通流线也十分单一,改变的可能性小。又由于它的承重方式是纵墙承重,很难在其中开辟出较大的空间供人集散。因此我在初期概念设计时着重解决这两个问题:交通和集散。

为了解决这两个问题,我们提出了两种方案:一是在外墙外侧增加一个走廊空间,用它串联起其他空间,作为核心的交通组织。这也是一些实际古建筑改造的项目中经常采用的一种方式,如大英博物馆,将其中庭作为主要的交通组织空间。二是在一端(与连廊相接的一端)增加一个大厅。最终将两者结合,形成了在外墙一侧增加一大一小两个方块的布局形式。

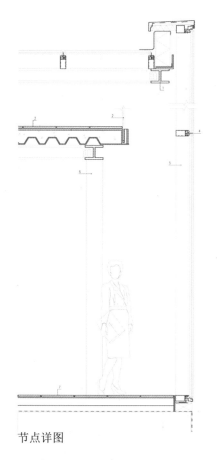

节点详图

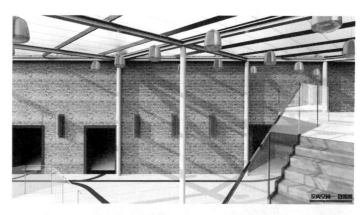

展厅室内效果图

扩建入口室内设计 | Interior Design of Enlarged Entrance

扩建入口室外效果图

扩建入口室内效果图

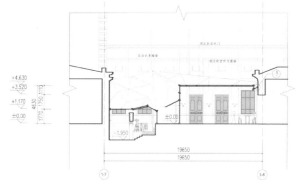

扩建入口室内 1—1 剖立面图

扩建入口室内 2—2 剖立面图

國粹賡續卷——歷史建築保護與再利用設計

答辯展示

创客书店室内设计 ｜ Interior Design of Maker Bookstore

创客书店室内效果图

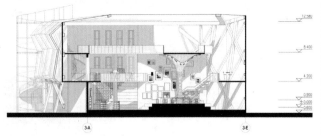

创客书店室内剖立面图

地下空间室内设计 ｜ Interior Design of Underground Space

地下空间室内效果图

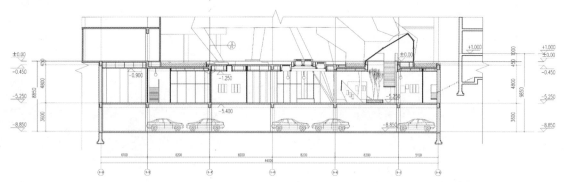

地下空间室内 1—1 剖立面图

导师点评

左琰
Zuo Yan

该组"文化寄居"将哈尔滨的一处文化建筑——哈尔滨文化馆作为研究对象，从这个历史保护建筑的新功能业态以及文化建筑与周边居民区的关系入手，力图将该建筑的历史文化价值挖掘出来，通过改扩建等手段实现文化宫与居民楼的连接和整合。总体设计逻辑性强，设计思路明确，凸显文化建筑自身的意义以及面向社区和公共的再生价值。有争议的是新加建的红色联系长廊，所谓新旧对话和融合是要把握好这个度，而本设计中夸张的造型和冰冷的材质对于这样有着宁静气场的历史建筑坊院不太搭调，并且文化宫室内也附加了一些金属折板作为反射区，冲淡了历史保护建筑固有的文化价值认同和历史原真性要求。

专家点评

叶铮
Ye Zheng

设计针对所选地块，很好地提出了"文化寄居"的总体概念，并表达了"寄"而不"混"的观点，较为贴切的应对了有百年历史的哈尔滨文化宫及周围地块的综合改造。

设计首先对整个地块场地进行改造，获得了良好的空间效果。同时以"新"围"旧"，建立起新旧的对话关系，赋予了整个区域以"文化寄居"的再生活力。而一体化的设计，从对文化宫建筑的改建入手，涵盖了从场地规划、建筑设计、室内改建、景观设计等方面，展示了学生全面的设计素养。其中对空间场地和新建建筑的设计，反映了学生良好的设计基础和专业才能。但在文化宫室内改建设计中显存不足，无论是设计手法、空间形象等方面，都缺乏进一步深入落地的设计表现，对新植入的设计与原有建筑的空间关系，处理得也较为生硬。

作为百年历史保护建筑的室内外改建，建议加强对建筑史料的考证和对比工作，并针对当下建筑的现况条件，从中梳理出保护与再建更新的具体内容。

This team's graduation project is entitled "Cultural Sojourn", which is a cultural building in Harbin—Harbin Cultural Center. The designer starts with the new features of this historic building as well as the relationship between this cultural building itself and the surrounding residential areas in an attempt to excavate the historical and cultural value of this building, and reconstruct or extend it to connect the cultural palace to the residential buildings. The overall design has a clear logic, and the design idea is explicit, highlighting the significance of the cultural building itself and its value to the community and the public. Controversially, the newly built connection hallway, which means a dialogue and fusion between the old and the new, should be handled properly. The exaggerated shape and cold materials in the design don't well match the tranquil historic courtyard. Moreover, some metal folded plates are used to make a reflecting area indoors, diluting the inherent cultural value identity and historical originality of the historic building.

For the target site, the design well proposes the overall concept of "cultural sojourn", and expresses a view that "sojourn" is not "chaos". This concept pertinently corresponds to the synthetic renovation of Harbin Cultural Palace, which has a history of one hundred years, and the surroundings.

First of all, the design renovates the whole site, achieving a good spatial effect. Meanwhile, the "old" is enclosed with the "new", building a dialogical relationship between the old and the new, giving the whole area a new life in the form of "cultural sojourn". As for the integrative design, it starts with the reconstruction of the cultural palace, covering many aspects such as site planning, architectural design, interior reconstruction and landscape design, revealing the designer's comprehensive design literacy. To be specific, the design of spaces and new buildings reflects the designer's solid design foundation and professional competence. However, there are shortcomings in the interior reconstruction design of the cultural palace: both design technique and spatial image lack practical design performance, and the spatial relationship between the new building and the old buildings is handled ponderously.

For the indoor and outdoor reconstruction of this time-honored historic building, I suggest strengthening research and compassion of the historical data of architecture, and making specific content of conservation, reconstruction and renewal according to the present conditions of the building.

山西夏门村梁氏古堡建筑群修复与再利用设计

Renovation and Reuse Design of Liang Family Castle Building Complex in Xiamen Village, Shanxi Province

高　　校：西安建筑科技大学
College: Xi'an University of Architecture and Technology
学　　生：崔维鹏、於天心、周文婷、马永强
Students: Cui Weipeng, Yv Tianxin
　　　　　Zhou Wenting, Ma Yongqiang
指导教师：刘晓军、王敏
Instructors: Liu Xiaojun, Wang Hui
课题评价：良好
Achievement: Good

学生感悟
Student's Thought

崔维鹏

青春是美好的，更是短暂的。我们不能在人生最辉煌的时刻虚度年华，碌碌无为。

迷人的彩虹出自大雨的洗礼，丰收的果实来自辛勤的耕耘。为此，我们应当好好地把握青春，去学那搏击风雨的海燕去展翅翱翔，学那高大挺拔的青松去傲雪经霜。

最后非常感谢此次参加"室内设计6+"，使我收获良多。

於天心

大学是一个新的环境、新的起点。在大学中的每一次认可都成为一次激励，我自己同样也期待下一次的认可，这样的期望，促使自己对待其他课程、工作时更加的努力。这种循序渐进的过程，使得自己改变了高中以前的不良习惯，习惯以高标准严要求来要求自己。

周文婷

夏门古村无无论从建筑结构还是从装饰艺术都体现着当地人民的智慧，以小见大，中华文化何尝不博大精深。此次对古村落设计与再利用机会，不仅教会我尊重传统艺术，重视地域艺术，而且让我意识到传承文化的最好方式就是赋予其时代活力，让其成为新的时尚。

马永强

四年沉淀，势必破茧，感谢在这毕业之际能与各位导师进行交流，让我收获更多，让我更加充实。无论是在学术上还是在今后的实际工作中，今天的努力与收获，都将成为我设计生涯中多彩丰富的一笔记忆。

场地介绍 | Site Introduction

地理区位 | Geographical Location

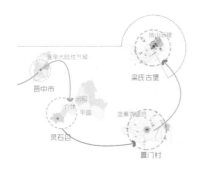

总平面图 | General Layout Plan

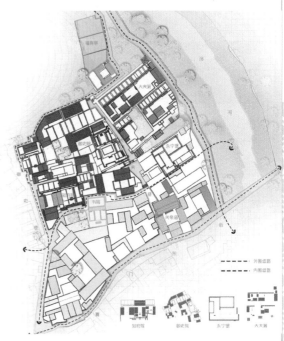

场地层级 | Site Hierarchy

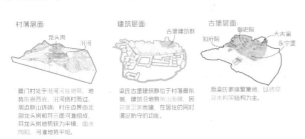

村落分析 | Village Analysis

夏门村基本概况：

夏门村古建分布：

夏门村路网分布：

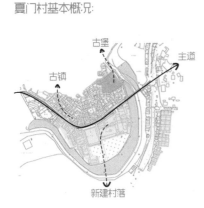

古堡分析 | Castle Analysis

古堡建筑群的外部联系：

古堡建筑群的构成：

古堡建筑群的内部联系：

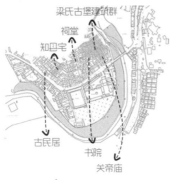
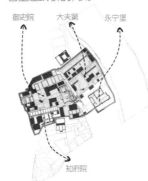

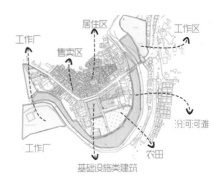

设计构思及改造 | Design Concept and Renovation

设计痛点 | Pain Spots of Design

通过调研走访以及和当地村民的交流，本场地最大的矛盾点便是村落日渐破旧的客观条件与当地居民向往乡土生活的无奈之间的矛盾。虽然除此之外，场地上还有大大小小的问题，如场地的防御体系弱、开发旅游产业难等，但相比而言前者更加可以体现出人们对乡土生活的向往。

NO.1: 古堡的历史
厦门村起源于隋唐时期的古战场，后随军事地位下降、商业发展，逐渐发展成村落。梁氏家族在明末清初迁入厦门村，作为当地最大的家族，在晋商文化的影响下，形成了"住守合一"的堡寨式建筑体系。

NO.2: 遗留的现状
梁氏古堡建筑群在清朝末年随着家族没落，被逐渐分割、买卖，至今仍有后人在此居住。当地建筑结构保存较为完好，建筑肌理、构件损坏较为严重，部分房屋、窑洞出现损毁。

NO.3: 文化元素
通过对村落的走访及调研，场地内对于竹编、木雕、竹帘、剪纸等文化元素的使用较为普遍。除此之外，场地内还有着大量的青石砖块、门牌、废弃磨盘等材料。在文化活动方面，厦门村是戏曲、泥塑、剪纸等文艺活动的特色村落。

NO.4: 所虑的人群
当下厦门村内所处人群以老年儿童为主，老年人对于厦门村的场地及历史了解较为深入，而年轻人群对于村落的了解则较为粗略。

NO.1: 美好的记忆
图内的两位老人回忆起儿时的记忆，眼睛里透露出无比喜悦的目光，显示出对往事古堡生活的向往。

NO.2: 无奈的乡愁
图为当地村民儿时照片与当地出版书籍上的历史资料上的对照，图内的牌坊现今已经不在，此刻村民眼中显露出对当下无奈目光和对原有生活的寄托。

NO.3: 悠久的乡土文化
厦门村的村口张贴厦门历史改编成的村歌，从中可以看出厦门村原有悠久的文化历史与浓韵的乡土文化。

设计说明 | Design Description

方案从尊重当地人群生活方式、尊重当地原有风俗文化传承的角度入手，激活当地传统文化脉络、重启当地文化历史遗迹，传承技艺、古为今用，改掉了一味追求经济效益而忽略当地生活的旅游规划。以"记忆场所·乡土情怀"为设计主题，表达当地村民对往日生活习俗的期盼和渴望。通过修复与改造，重新激活当地原有的生活场所，提高当地人居环境，使悠久、醇厚的乡土文化可以源远流长。

设计构思 | Design Concept

设计原则 | Design Principles

修旧如旧、依旧修旧

1. 原材料修复

采用当地原材料砖石、泥土等土生材料进行修复。

2. 原方式修复

尊重场地上个建筑结构及方式的运用，如：窑洞建筑拱券的样式、原有构架形式，以及现原有的建筑技艺。

3. 原功能再现

对场地内的原有功能尽量保留，特别对于一些标志性建筑的使用功能定位。

新旧对比、古为今用

1. 新材质修复

将新材料与原有材质肌理做对比，从而体现历史痕迹。

2. 新物件修复

在村民生活区等类似区域，为了未来更好地提高人的居住生活体验，采用传统元素/建筑与现代家居相结合，创作出古为今用的修复改造模式。

3. 新方式修复

使用传统材质及元素，使用新的方式及手法进行创作，满足在继承文化的基础上推陈出新。

尊重历史、保护现状

1. 临时性设施

对于已破损的建筑，可以在其原有的建筑结构上，做临时性的局部改造，从而在保护的前提下，增加其使用功能。

2. 减少干预

对有着较大历史价值的区域，采取最小干预原则，也就是不做更改，保留原样将其作为保护区展示，供人群观赏，铭记历史。

古堡建筑分析 | Castle Building Analysis

御史院 | Censor Courtyard

御史院原为梁氏族人居住的地方，有九户一堂（祭祖堂），至今仍有六户人家在此居住，当下建筑院落空间被分割，打破了原有的院落格局，御史院内建筑保护较为完善。

大夫第 | Manor Garden

大夫第内部多为砖砌窑洞及砖木梁构架建筑，有着形成了"堡中堡、窑中窑"的建筑形式，用于防御。而今大夫第被毁严重，往日建筑布局形式早已不在，部分院落被毁。

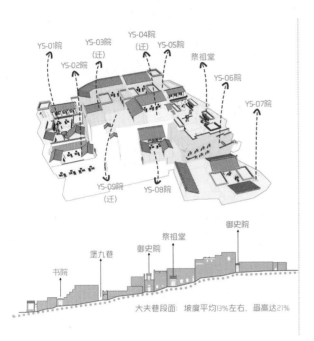

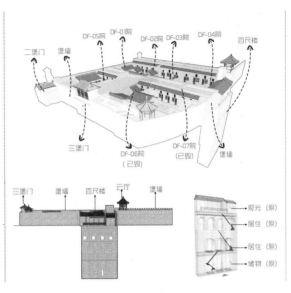

永宁堡 | Yongning Castle

永宁堡院落具有很强的防御功能，是典型的堡寨式民居。建筑为多层衬窑结构。院内有直接通往其他院落的暗道，是梁氏古堡一个极为重要的出入口。

知府宅 | Prefecture Magistrate Courtyard

知府院院落群位于夏门村中部，西邻御史巷，南接堡九巷，东有大夫巷。共有三个院落，为砖拱窑洞结构与砖木梁架结构相结合的院落群。

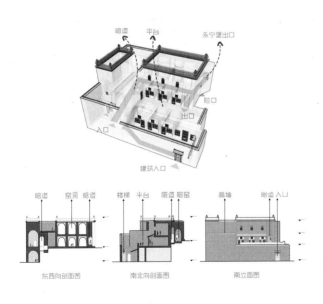

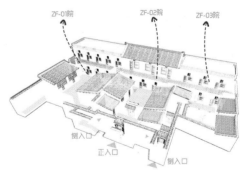

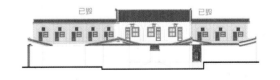

建筑改造与效果表达 | Building Renovation and Effect Presentation

御史院 | Censor Courtyard

御史院是原有梁氏家族的家族聚居地，位于夏门村北部，西邻御史巷，东接大夫巷，北至后堡道；御史院共有九个院落，外加一座祭祖堂，被称为"九户一堂"，至今仍有六户在居，三户外迁。当下建筑院落空间被分割，打破了原有的院落格局，但建筑保护较为完善，使用功能定位仍以居住为主。

在改造时遵循新旧对比的改造原则，以提高当地人的人居生活体验。

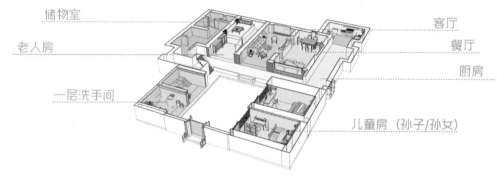

一层平面图

二层平面图

效果图

大夫第 | Manor Garden

大夫第原为梁氏家族内的会客及活动区，也用于住守防御防守之用，在当下作为夏门村的文化技艺传承地及文化活动区使用。

在改造时遵循依旧修旧的改造原则、最小干涉的保护原则。

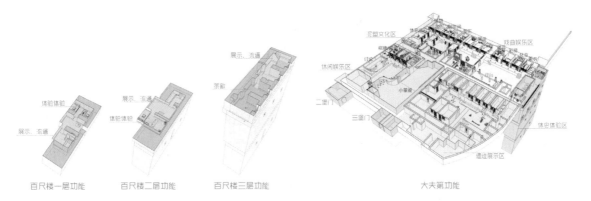

效果图

永宁堡 | Yongning Castle

永宁堡是原有的古堡防御类建筑，空间独特，在当下在不需要防御措施的情况下，我们将其改造成为传承夏门村历史文化记忆的记忆展示馆，用以传承历史记忆，继承乡土文化。

在改造时遵循依旧修旧、新旧对比的保护原则。

一层功能图　　　　　　　　　　　二层功能图

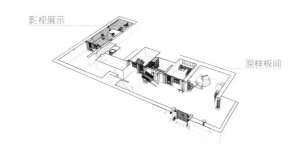
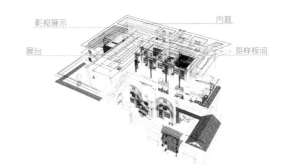

知府宅 | Prefecture Magistrate Courtyard

知府院院落群位于夏门村中部，西邻御史巷，南接堡九巷，东有大夫巷。共有三个院落，为砖拱窑洞结构与砖木梁架结构相结合的院落群。

在改造时遵循新旧对比的保护原则，保护现场建筑及场地构件，以现代材料手段进行修复，满足当下功能使用。

二层功能图

一层功能图

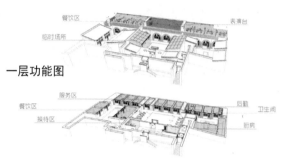

效果图

人群活动分析

方案通过以当地人群需求向往乡土生活的情怀为切入点，在对夏门梁氏古堡的功能及其环境等方面进行更改后，我们对其不同年龄阶段的人群一天的生活业态进行分析，发现其一天在古堡中的生活轨迹，可以形成一个完美的循环，从而体现出古堡的情感圈层关系。

鸟瞰图

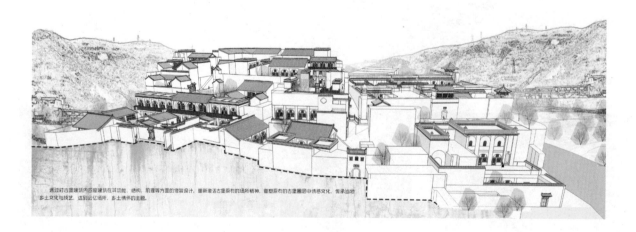

导师点评

朱飞 Zhu Fei

西安建筑科技大学的山西夏门村梁氏古堡建筑群修复与再利用设计项目，体现了过程的系统性、目标设定的逻辑性、方案表现的完整性，以及技术应用的合理性。该方案涉及到御史院、大夫第、永宁堡和知府院四大区域，把居住、餐饮、文化活动体验、专题展馆等功能很好地融合到不同的空间区域之中，对传统建筑采用了诸如"修旧如旧""新旧结合"等不同的修缮、改造手法，满足了不同的使用要求。该方案的闪光点还在于同学们始终关注当地居民的现实生活，重视本土传统文化的传承，这使得整个方案有了一条可持续的生命线。

方案可以在此基础上，进一步在室内设计、陈设设计等方面进行深化，利用形态、色彩等要素彰显传统文化的魅力。

Xi'an University of Architecture and Technology's design project, namely the renovation and reuse of the Liang family castle building complex in Xiamen Village, Shanxi Province, reflects the systematization of the design process, the logicality of the objective setting, the integrity of the scheme and the rationality of the technical application. The scheme involves four areas: the Censor Courtyard, Manor Garden, Yongning Castle and Prefectural Magistrate Courtyard. Living, catering and cultural experience, special exhibition and other functions are well integrated into different spatial areas, and the traditional buildings are renovated with "the old still looking old" or "some old looking new", in order to meet different requests for utilization. One highlight of the scheme is that the students are always concerned about the local residents' life and focus on inheriting the local traditional culture. That's how the scheme is given a sustainable lifeline. The scheme can, on that basis, be further deepened in terms of interior design and furnishing design, to reveal the charm of the traditional culture with forms and colors.

专家点评

刘磊 Liu Lei

首先整个设计很好地对应了人文主题，符合设计为人服务的宗旨。设计逻辑清晰，整个设计具有一定的实用性。

（1）古城建筑群的修复再利用从历史人文入手，有很好的设计切入点，对于场地现状的罗列，并作了很好的总结。设计理念来源于人群需求，贴合人文设计理念。

（2）设计分析图表达到位，从内到外，从外到内的设计表达分析，设计逻辑性相当清晰。

（3）从具体的设计内容来说，也是具有自己的优势所在的，比如古堡建筑内部的空间功能划分，就能根据使用者的具体情况而定，空间的划分使用是具有一定的设计实用性。所以跟"记忆场所—乡土情怀"主题来说，还是有一定的切题度的。

（4）版面整齐有序，版面有一定的设计讲解性，整体色调方面也是具有一定的统一性，统一的灰绿色调子。但美中不足的是，在过于追求色调统一的过程中，忽略了同色调中色彩层次的细节变化。整体色调也就略显单薄。

总的来说，设计尊重了当地人的生活方式，从当地人的生活入手，重新激活了当地原有生活状态，对于乡土改造类设计都有一定的借鉴意义。

First of all, the whole design well corresponds to humanity, aligned with the aim of design, which is to serve people. The design logic is clear. The entire design has good practical applicability.

(1)The ancient architectural complex is renovated and reused from the perspective of historical humanity, which is a good entry point. The details of the site context are well summarized. The design idea comes from human demand, and fits the concept of humanistic design.

(2)The analysis chart provides effective information as a design expression analysis from inside out and from outside in. The design logic is fairly clear.

(3) In terms of specific design content, the design group has its own strengths. For instance, the interior spatial functions of the castle can be partitioned as the case may be, and the spatial partitioning has certain practical applicability. So, it is relevant to the title "A Place in Memory—Country Affection".

(4) The page layout is in good order and able to indicate the content of design. The overall hue is greyish green, which looks harmonious. However, the only drawback is that the slight change in color gradation is ignored due to the excessive harmony of hue. So the overall hue is slightly faint.

On the whole, the design shows respect for the local people's lifestyle, and starts with the local people's life, reactivating the local original living conditions. This is of reference significance to the renovation design of rural buildings.

恭王府博物馆（忻州工作站）展览设计
· 晋作家具制作技艺精品展

Exhibition Design for Prince Kung's Museum (Xinzhou Workstation)
—Exhibition on Shanxi-style Furniture Craftsmanship and Masterpieces

高　　校：北京建筑大学
College: Beijing University of Civil Engineering and Architecture
学　　生：刘宇麒、闫铮、曹志伟
Students: Liu Yuqi, Yan Zheng, Cao Zhiwei
指导教师：杨琳、陈静勇
Instructors: Yang Lin, Chen Jingyong
课题评价：良好
Achievement: Good

学生感悟
Student's Thought

刘宇麒

　　经过这一个多学期的努力，我们的毕业设计终于完成。毕业设计对于我们来说，既是一次小小的挑战，又是对我们大学几年所学知识的测验。
　　在做毕业设计的过程中，我们遇到了很多问题，当我们遇到难题时，在经过陈静勇老师、杨琳老师的帮助下，这些难题得以解决，设计也能顺利的完成。所以我非常感谢这两位老师为我们提供的帮助。

闫铮

　　毕业设计，是我们大学里的最后一道大题，只要认真对待，所有的问题也就迎刃而解。我必须有扎实的理论功底和丰富的实践经验才有可能保质保量地完成预定的设计目标。回顾四年以来的专业学习，自己还是存在一些不足和遗憾之处，但从整体上来看，经过自己不懈的努力还是取得了长足的进步，给我大学的生活和学习划上了较为满意的句号。

曹志伟

　　毕业设计是一次意志的磨练，也是对我实际能力的一次提升，也会对我未来的学习和工作有很大的帮助。非常感谢我的导师杨琳老师和陈静勇老师的悉心指导，让毕业设计圆满结束。

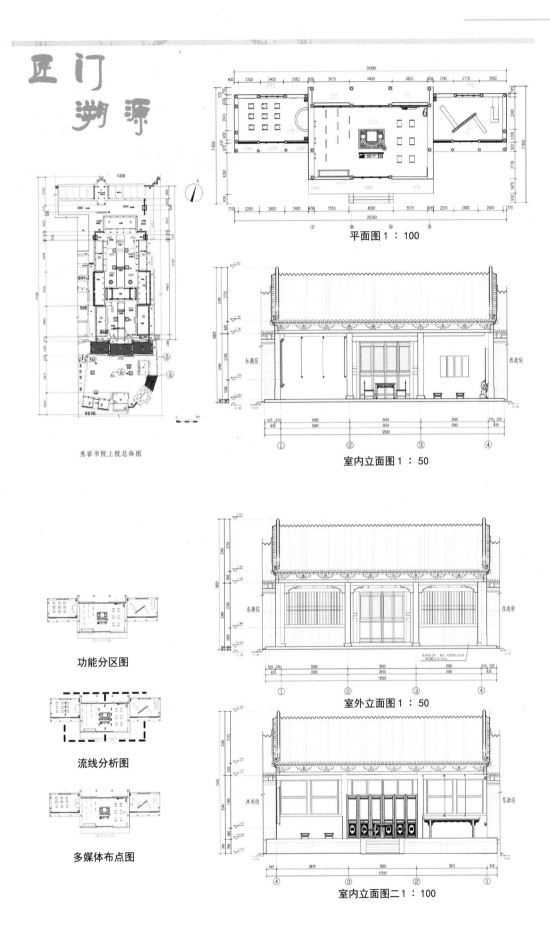

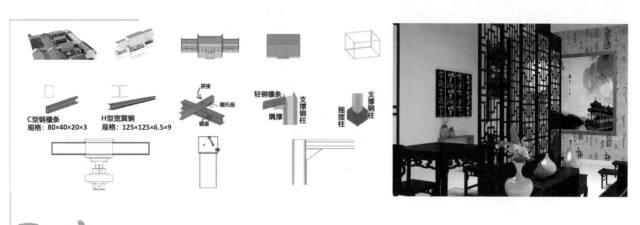

匠心独运

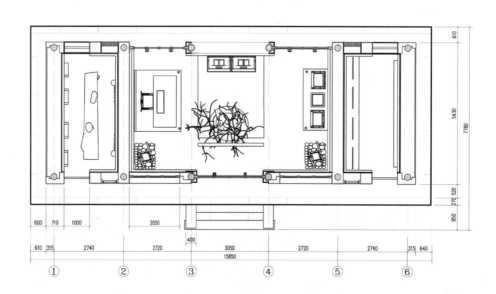

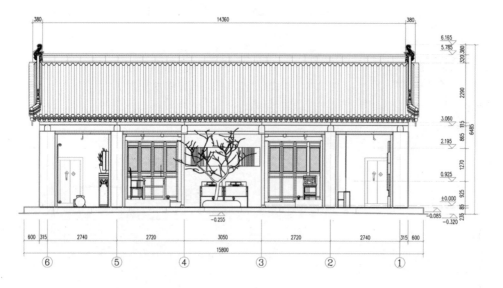

 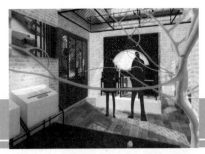

仪銮殿主展厅约为 45.9 ㎡，主要进行书院场景复原以及传承人手稿。展厅中间布置了一棵树。实际的木材的存在会让人有更好的身临其境的感觉。树的主体成白色。观展预计 10～20 分钟。

仪銮殿左右展厅面积约为 15.9 ㎡，主要进行家具的历史脉络，以及各种家具的榫卯纹样的展示，以视频的方式展示榫卯插接的精妙之处。观察预计每个展厅 30～50 分钟。

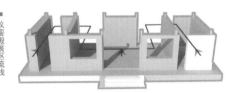 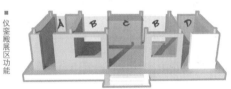

- 仪銮殿展区流线
- 仪銮殿展区功能

A 晋作家具发展史展厅
B 书院场景复原区域
C 传承人手稿展示区域
D 晋作家具榫卯纹样展厅

國粹賡續卷——歷史建築保護與再利用設計

答辯展示

匠人材器

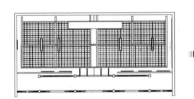 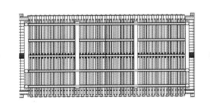 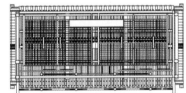

匠人材器天花图

展厅后面区域是互动区域。由于观看展览的人可能会比较多，所以在这里设置了一面镜子，在有人操作的时候，站在远处的看不到具体情境的人也可以通过镜子的反射观看他人雕刻的过程。

展厅东侧，运用晋作家具对明代的会客房进行了一个场景还原，同时还能展示忻州当地的非遗项目。操作台的两侧放有工具的展品，中间的操作台是互动区域的中心，工人们自己动手对木头进行加工等操作。

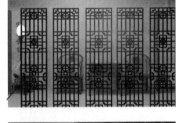

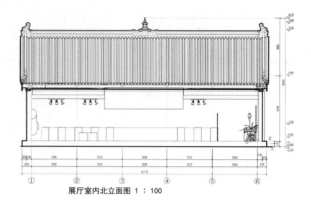

展厅室内正立面图 1：100

展厅室内北立面图 1：100

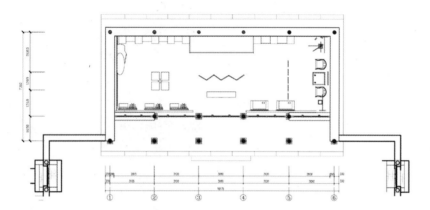

展厅平面图 1：100

古建筑的保护

悬挂的展示方式需要挂点，在不破坏原有建筑的前提下，采用了木框架的结构形式——在原有的梁架结构上，搭建具有一定承载能力的木条，连接上使用榫卯结构，没有任何一根钉子的影子，这样的做法最大限度地保护了秀容书院古建筑。

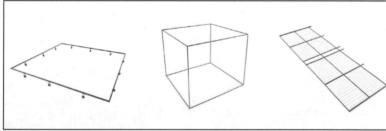

宣传手册和周边环境设计

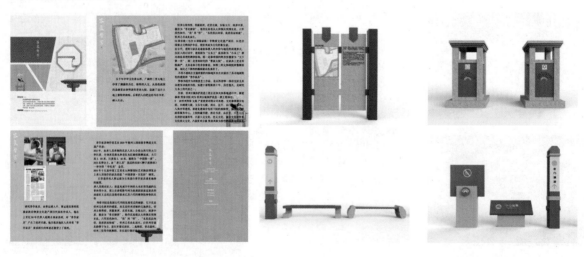

导师点评

兆晖 Zhao Hui

该方案对秀容书院及周边环境进行了深入细致的调研和资料梳理，并详细地规划了书院的功能、流线。清式两庭两进的书院中院层次第落，提升了展示空间的景深感，中庭的布局温馨舒适，带有现代元素的复古风室内装饰与古典家具展示相得益彰，为诠释晋式家具遗风搭建了一个穿越古今的平台。以明式晋作家具作为核心切入点，将晋作家具传统工艺、历史、实物展现及应用，用现代的演绎方式展示给参观者一个完整的晋作家具编年史。室内设计的简中设计元素既表达出对历史建筑的尊重，又加强了展陈空间的带入感，予参观者以更深刻的全方位体验。

The project conducts an in-depth research on Xiurong Academy and its surroundings, and makes detailed planning for its function and flow line. The academy is a Qing-style two-courtyard building, which is well-spaced and provides a strong sense of depth of field. The atrium is furnished cozily with modern-style vintage interior ornaments and classic furniture that bring out the best in each other, building an ancient and modern platform for interpretation of the legacy of Shanxi-style furniture. The complete annals of Shanxi-style furniture, including the traditional craft, history, real objects and application of Shanxi-style furniture, especially the Shanxi-style furniture of the Ming Dynasty, is shown to visitors in a modern way. The simple Chinese-style elements in the interior ornaments show respect for the historic building, and enhance visitors' sense of integration into the display space, giving them a deeper all-round experience.

专家点评

陈静勇 Chen Jingyong

山西忻州秀容书院始建于清乾隆四十年（1775年），当时忻县称秀容县，故以此得名，为忻州市第一所学府。清光绪二十八年（1902年），改称"新兴学堂"，创山西书院改学堂之首例。2004年6月10日由山西省人民政府公布为省级重点文物保护单位。目前是作为国家文化和旅游部设立的山西省国家非物质文化遗产保护与研究工作站。

以秀容书院作为联合毕业设计"国粹赓续——历史建筑保护与再利用设计"选点之一，体现出课题特色。

该方案以"容里晋书：晋作家具传统制作技艺精品展"展陈设计为题，旨在探讨工作站在秀容书院保护与利用中的建设模式。方案设计以历史建筑保护和传统家具制作技艺谱系调研为基础，突出晋作家具传统制作技艺所遵循的非物质文化遗产要走"见人见物见生活"的活态保护与传承之路原则。

Shanxi Xinzhou Xiurong Academy was founded in the 40th year of the reign of Emperor Qianlong of the Qing Dynasty, when Xin County was called Xiurong County, and is the first academy of Xinzhou. In the 28th year of Emperor Guangxu of the Qing Dynasty (1902), it was renamed "Xinxing School", and of the academies in Shanxi, it was the first to be transformed into a school. On June 10th, 2004, it was identified as a provincial key historical and cultural site under government protection by Shanxi Provincial People's Government. Currently, it is a national intangible cultural heritage conservation and research center established by the Ministry of Culture and Tourism in Shanxi Province.

Xiurong Academy is a characteristic topic for joint graduation design National Quintessence Continuation—Historic Building Conservation and Reuse Design.

The project is titled "Rongli Book of Shanxi: Exhibition on Shanxi-style Furniture Craftsmanship and Masterpieces", aimed at developing a building model for the conservation and utilization of Xiurong Academy. The design is based on historic building conservation and the pedigree of traditional furniture craftsmanship, highlighting the principle of active protection and inheritance that "there must be inheritors and real objects in real life" followed by Shanxi-style furniture craftsmanship.

万花筒——民国时期人物展
Kaleidoscope——Personages of the Republican Period

高　　校：南京艺术学院
College: Nanjing University of the Arts
学　　生：马悠庭　范聪达　孙昱　高金宇
Students: Ma Youting, Fan Congda, Sun Yu, Gao Jinyu
指导教师：朱飞
Instructors: Zhu Fei
课题评价：良好
Achievement: Good

学生感悟
Student's Thought

马悠庭

参加"室内设计6+"活动以来，成长了很多，得到了知名高校老师的指导，交到了五湖四海的朋友，学到了很多知识，希望本届活动的结束是我们大家感情的开始。

范聪达

感谢"室内设计6+"，让我增加了专业知识和水平，锻炼了自己，得到了许多高校老师的熏陶，也认识了全国各地的朋友。同时我也很感谢朱飞老师对于我们耐心的指导，我也很感谢我的同伴们不懈的一起努力。再次感谢"室内设计6+"！

孙昱

非常荣幸能够参加"室内设计6+"这个活动，在此活动中，最开心的是与小伙伴们一起努力的日日夜夜。在比赛中，很荣幸能够与全国各地老师和同学一起学习。不论结局如何，在团队协作以及老师对我们学业和心态上的指导，都使我受益匪浅。

高金宇

感谢在南京艺术学院这四年，我们收获了友情、理解、青春、乐观和团结。在这个大家庭里面我们互相学习交流和进步，经过朱老师的教导和组员们的不懈努力，我们最终完成了这个毕业作品，为大学生活画上了一个圆满的句号，谢谢大家。

基地分析 | Base Analysis

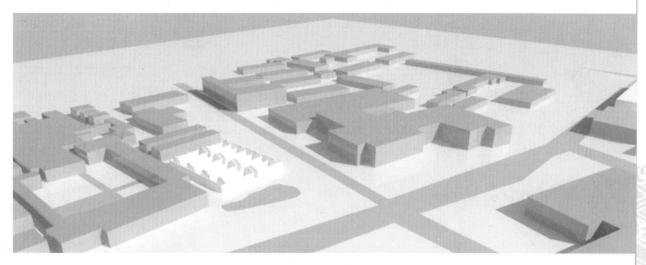

南京总统府位于南京市玄武区长江路292号，既有中国古代传统的江南园林，也是近代时期的建筑遗存，是中国近代建筑遗存中规模最大、保存最完整的建筑群。

南京总统府东朝房为清代建筑。清朝时为两江总督署吏、户、礼科房。太平天国时为官员等候天王接见的地方。后为总统府警卫团兵舍。如今则改造成了展览场所。

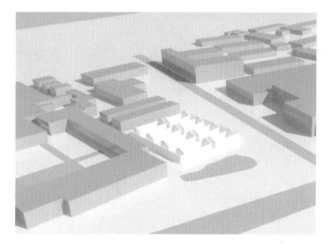

场地选址

人流走向

周边建筑

交通概况

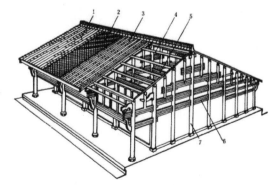

穿斗式构架是中国古代建筑木构架的一种形式，这种构架以柱直接承檩，没有梁，原作穿兜架，后简化为"穿逗架"。穿斗式构架中柱承檩的作法，已有悠久的历史。

特点：穿斗式构架用料较少，建造时先在地面上拼装成整榀屋架，然后竖立起来，具有省工、省料、便于施工和经济的优点。同时，密列的立柱也便于安装壁板和筑夹泥墙。因此，在中国长江中下游各省，保留了大量明清时代采用穿斗式构架的民居。

设计概念分析 | Design Concept Analysis

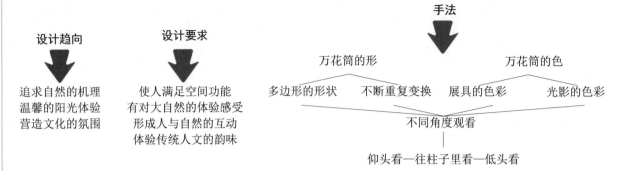

设计趋向
追求自然的机理
温馨的阳光体验
营造文化的氛围

设计要求
使人满足空间功能
有对大自然的体验感受
形成人与自然的互动
体验传统人文的韵味

手法
万花筒的形：多边形的形状、不断重复变换
万花筒的色：展具的色彩、光影的色彩
不同角度观看
仰头看—往柱子里看—低头看

关系分析 | Relation Analysis

新与旧
改造与保护
更新于保留

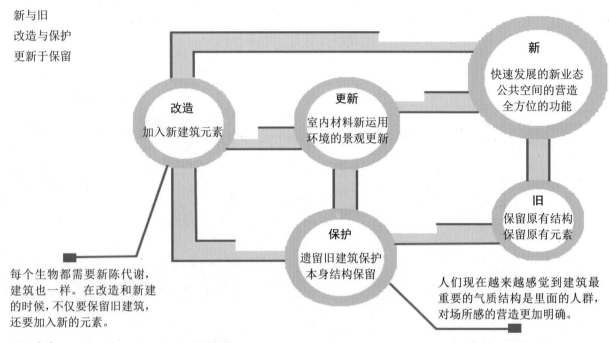

改造 加入新建筑元素
更新 室内材料新运用 环境的景观更新
新 快速发展的新业态 公共空间的营造 全方位的功能
保护 遗留旧建筑保护 本身结构保留
旧 保留原有结构 保留原有元素

每个生物都需要新陈代谢，建筑也一样。在改造和新建的时候，不仅要保留旧建筑，还要加入新的元素。

人们现在越来越感觉到建筑最重要的气质结构是里面的人群，对场所感的营造更加明确。

■ 态度
尽量去还原原本空间的特质，呈现建筑本身的痕迹，通过保护性改造，更新利用让空间得以升级，环境还原了以前的样子，但是有的空间却有了新的面貌，让人们对总统府的印象也有新的印象。

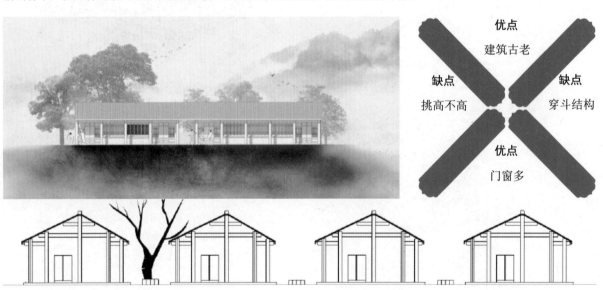

优点 建筑古老
缺点 挑高不高
缺点 穿斗结构
优点 门窗多

平面布局 | Plane Layout

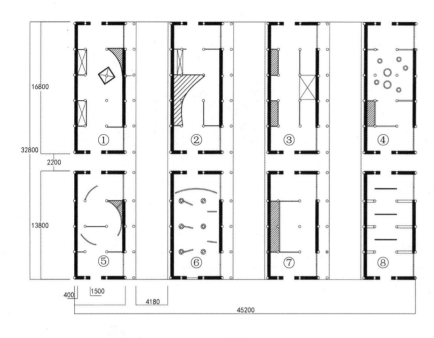

① 文人展厅 "真名士、自风流"
目的：展现民国拥有不同思想的文人

② 科学家展厅 "洒脱与落寞"
目的：展现为了国家复兴的科学家

③ 明星展厅 "千载一时"
目的：展现独具特色的民国范的影星

④ 戏曲展厅 "传奇"
目的：展现戏曲家在民国独有的魅力

⑤ 教育家展厅 "师道的重量"
目的：展现民国以书为伍的教育现状

⑥ 画家展厅 "旧时风月"
目的：展现不同风格画家的人物故事

⑦ 军人展厅 "气节长存"
目的：展现民国战乱为国捐躯的军人

⑧ 政治家展厅 "悠悠水长"
目的：展现无路探径寻国运的政客

项目目标
- 运用博物馆藏品、知识背景以及恰当的展示手段，展示丰富的民国文化。
- 尽力向所有观众展示现有文物的内涵和外延。
- 吸引观众参观展览，特别要鼓励16～35岁的年轻人。
- 实现展览项目中的关键信息。
- 在万花筒的形式感与美感上与民国文化展览相结合，由不同的人物职业组成一个绚丽的万花筒。

关键信息
- 运用博物馆藏品、知识背景以及恰当的展示手段，展示丰富的民国文化。
- 总统府内设有民国时期人物展览。
- 展览能够调动观众的五感，且内容丰富。
- 观众能从展览中感受到民国范儿。
- 观众被激起更多兴趣了解民国文化。
- 应引导观众思考民国韵味如何在当代全球化环境下进行继承与创新。

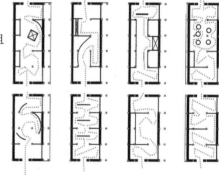

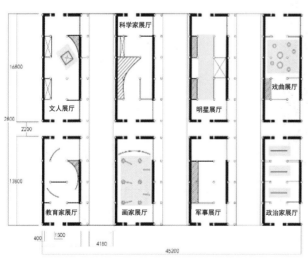

多媒体区域

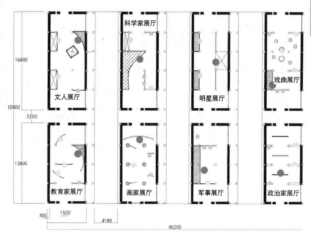

● 一级展板
　 二级展板
　 三级展板

光线分析 | Light Analysis

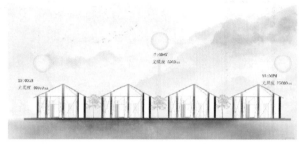

古建利用 | Reuse of Ancient Buildings

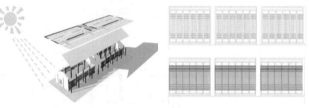

平面布局从建筑的光照强度出发,在光照强的一方,采用自然光补光的方式,尽量保证原建筑本身的古韵感,在尊重建筑风格的方式上进行改造与展览。

互动装置 | Interactive Installation

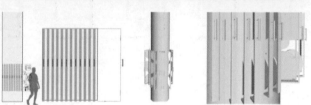

设计说明 | Design Description

画家展厅以互动展厅为定位，在展厅中运用手机做为媒介，可以站在具体位置进行扫描，从手机中得到展示讯息，同时在展厅柱子上的装置，在装置内有展示介绍与影视资料介绍，展馆内容是民国时期画家，而通过影视的画面传达，更能直观的展现其画家形象与作品风貌，而互动元素的出现也给展厅加入了趣味性。

科学家展厅出发点是用民国时期的装饰图案例如西式拱门与简化的装饰线条，加上民国时期浑厚庄严的建筑体量感和简洁区块分割的特点，我们在效果图中采用了手绘的形式，用线条来体现其展厅中展具的体块感及展厅的简洁形式。

戏曲展厅主要运用灯光效果与光照效果相结合，展厅效果主要侧重于万花筒的色彩运用，从展馆内容出发，主要展示的是京剧代表人物梅兰芳，抓住戏曲感的中国风为主要特点和展示形式相融合进行展示。

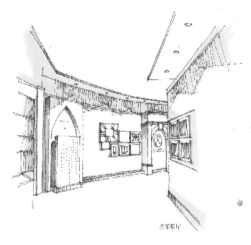

展面设计 | Display Panel Design

明星展厅是用万花筒的形状来展现空间,从建筑本身出发,以做纵深感空间为主要效果,运用地面的反射感给参观者一个眼花缭乱的感觉,同时以展厅内容——民国影星出发,抓住其绚烂星光特点,来展示空间万花筒效果,给参观者一个"黑屋子"的场景效果。

军人展厅内容是展示在民国时期抵御外辱牺牲的士兵。用场景还原的形式,营造一个庄严肃穆的战斗场景,使整个展厅给参观者悲壮感。

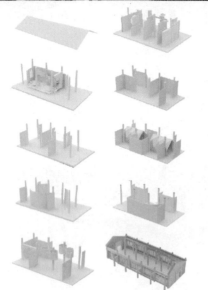

政治家展厅是以民国装饰色彩为设计出发点,用青色与红色相呼应,通过色彩的运用给展厅以视觉上的民国风。整个展厅分为四个部分,介绍民国时期主要的四位政治人物。展厅中用屏风来隔绝不同的区域,既划分了展厅同时也展现民国时期的政治人物的换代。

导师点评

林怡
Lin Yi

该设计以南京总统府东朝房这一清代中式传统建筑为设计对象，分析了穿斗建筑作为展览建筑的优势与缺陷，进而针对室内立柱较密集的特点，巧妙地将展陈内容与立柱相结合，使得立柱亦成为展览构成要素之一，并形成独具特色的展陈空间，与展示主题的时间线索"民国时期"形成了较好地回应。通过对柱、檩、窗等原有建筑构件的保留和再利用，让置身其中的参观者可快速融入新旧掺杂、中西碰撞的民国时代氛围，更充分地体会和感知所展示内容。值得肯定的是，该设计考虑了自然光渗透对展陈平面布局影响，但对于直射阳光可能带来的眩光干扰问题缺少应对措施，建议可从窗户的玻璃材质、人工照明的平衡等角度进行展陈空间光环境的改善。

The design is focused on an east-facing house in the Nanjing Presidential Palace, a Chinese traditional building built in the Qing Dynasty. The designer analyzes the advantages and disadvantages of a building of column-and-tie structure as an exhibition building, and then combines display content with the columns skillfully considering that there are a lot of columns indoors, making the columns a component of the exhibition, forming a unique display space, which well matches the time cue of the display theme—the "Republican Period". With the original building components, such as columns, purlins and windows, retained and reused, visitors in the interior space can quickly blend themselves into the old and new Republican atmosphere of Chinese and Western styles, to better experience and perceive the display content. On the positive side, the design considers the influence of natural light on display layout, but it doesn't put forward countermeasures against dizziness caused by direct sunlight. I suggest improving the lighting environment in the display space by changing the window glass or adopting artificial lighting.

专家点评

姚丽
Yao Li

南京艺术学院的"民国万花筒"传统建筑改造展示设计方案，选址中国近代建筑遗存中规模最大、保存最完整的建筑群——南京"总统府"，并以民国人物为展览内容。方案策划富有创意，体现了传统建筑和展览内容的高度统一，展现了丰富多彩、名人荟萃的民国人物风貌。作品主要从形、色、光三个方面来表现"万花筒"这一个主题。同学们利用万花筒所产生的多边形、折射、镜像等形态特征，设计了一个贯穿展览的"母体"；很好地使用了原有建筑序列柱体的红色，以及万花筒所特有的对比色关系；方案还较好地考虑了自然光线和人工照明之间的相互联系，使得观众在参观传统建筑的同时，能够较好地观赏具有现代技术应用的展览。

该方案在内容策划和形式设计上还需进一步深入，要处理好人物选择的代表性和普遍性之间的关系，以及商业展览和较为严肃、严谨的主题展览之间的关系。

Nanjing University of the Arts' graduation project is titled "Kaleidoscope of the Republican Period" concerning the renovation of a traditional building. The design object is the Nanjing "Presidential Palace", the largest and best-preserved building complex of modern Chinese buildings, and the personages of the Republican Period are displayed as well. The creative design reflects the high unity between the traditional buildings and the display content, and reveals the colorful look of the Republican Period, when there were many celebrities. In the work, the theme "kaleidoscope" is expressed from the perspective of form, color and light. The designer designs a "parent body" that appears throughout the exhibition according to the morphological characteristics of the kaleidoscope including polygon, refraction and mirror reflection; good use is made of red, the color of the columns in the buildings, and the contrast color, unique to the kaleidoscope; also, the design well considers the relation between daylighting and artificial lighting, so that viewers could enjoy the sight of an exhibition held by modern technology while visiting the traditional buildings.

In terms of content planning and form design, the project needs improving. The relationship between the representativeness and universality of personages selected, as well as the relationship between a business show and a serious theme, needs to be well handled.

合·游——浙江省薛下庄村"二十四间"老建筑保护与再利用设计研究

Heyou—Conservation and Reuse Design of "Twenty-four" Old Buildings in Xuexia Village, Zhejiang Province

高　　校：浙江工业大学
College: Zhejiang University Of Technology
学　　生：章佳祺、章家骐
Students: Zhang Jiaqi, Zhang Jiaqi
指导教师：吕勤智、王一涵
Instructors: Lv Qinzhi, Wang Yihan
课题评价：良好
Achievement: Good

学生感悟
Student's Thought

章佳祺

通过这次的比赛，我们慢慢学会更加理性与有逻辑地去发现问题、解决问题。秉持"以人为本"的原则去设计，更多考虑村民的诉求。设计过程中，我们的指导老师也给到了我们巨大的帮助，让我更多地了解到设计应有的思考与过程，这也是这次比赛带给我的最大的收获。

章家骐

参加这次竞赛实际上是由于阴差阳错的选择，但是在这次毕设的过程中无数次地感慨之前的无心之举让自己收获很多。这次的设计深入到了自己此前本科阶段从未涉及的深度和细节，拓展了自己的知识边界和操作能力，不论是经历还是结果，都是本科阶段的一个理想结尾。

"二十四间"建筑概况 | Overview of Twenty-four Houses

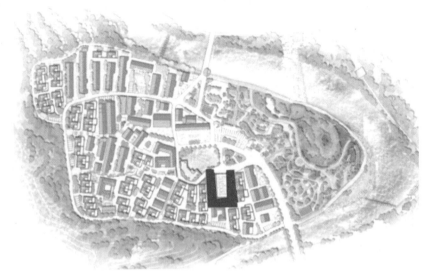

"二十四间"位于薛下庄村的中心位置，北临莲塘，故村民自称"莲塘薛氏"，村内建筑主要种类有木架构、砖木及砖混建筑，建筑面貌比较多样。

基地分析

视线分析

保留建筑

新建建筑

原始建筑

浦江县地处名胜风景区腹地，周边高品质资源云集。北联桐庐、诸暨，打包发展山水旅游联动区，实现客源共融，共生发展。南临义乌、金华，充分利用其文化旅游资源优势亏补，实现客流共享。县域周边乡村旅游联动态势基本形成。

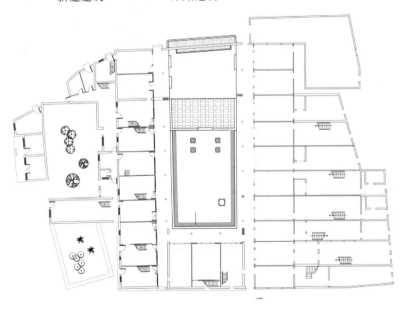
原始平面图

设计概念分析 | Design Concept Analysis

传统合院的内在价值 | The Intrinsic Value of the Traditional Courtyard

合院形式形成的内向空间使合院成为了一个高频使用的交通空间和活动空间。

"二十四间"的空间发展 | The Spatial Development of Twenty-four Houses

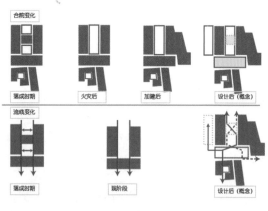

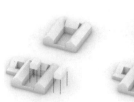 

原有建筑　　新增合院体系　　组合式院落形成

新增活动空间　　多样化合院体系　　强调入口形成方案

在原有合院空间的基础上，加建了新的活动空间，合理利用原有建筑形式，且不对原有建筑造成破坏与影响。

总平面图 | General Layout Plan

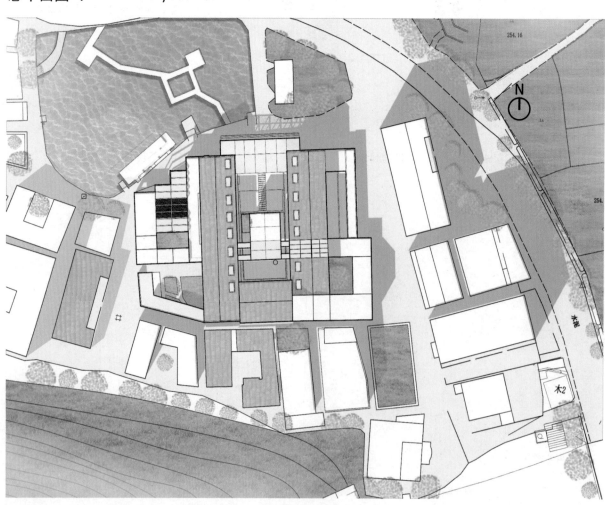

效果图 | Rendering

村落公共合院

公共合院部分主要作为"二十四间"建筑群落的观景展览部分使用。

村落公共合院 2F

村落公共合院 1F

建筑设计分析 | Architectural Design Analysis

民俗体验馆建筑 | Folk Custom Experience Buildings

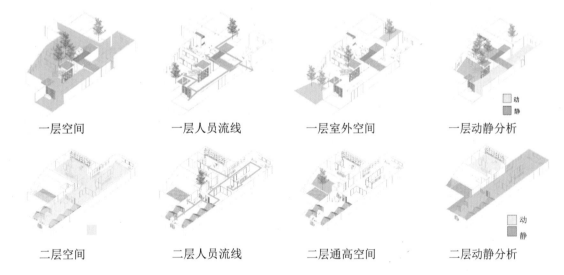

一层空间　　一层人员流线　　一层室外空间　　一层动静分析

二层空间　　二层人员流线　　二层通高空间　　二层动静分析

花园餐厅 | Garden Restaurant

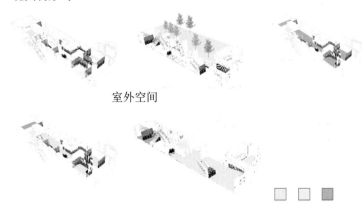

室外空间　　　　　　　动静分区

使用频率　　　　　　　高 中 低

民俗体验馆的概念是根据浦江县当地的文化特色：浦江书画、竹木根雕等文化背景得到的，主要功能以体验展示为主。

花园餐厅部分响应浦江县政府号召的"花漫浦江"概念，加上"二十四间"所处的薛下庄村以农业为主要产业的特征，故设计花园餐厅作为对外旅游的一个重要空间组成部分。

莲塘书院建筑分析 | Analysis of Buildings in Liantang Academy

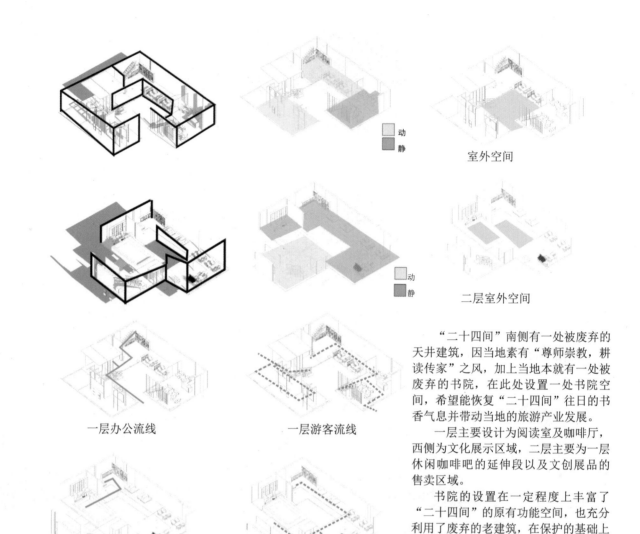

室外空间

二层室外空间

一层办公流线

一层游客流线

二层办公流线

二层游客流线

"二十四间"南侧有一处被废弃的天井建筑，因当地素有"尊师崇教，耕读传家"之风，加上当地本就有一处被废弃的书院，在此处设置一处书院空间，希望能恢复"二十四间"往日的书香气息并带动当地的旅游产业发展。

一层主要设计为阅读室及咖啡厅，西侧为文化展示区域，二层主要为一层休闲咖啡吧的延伸段以及文创展品的售卖区域。

书院的设置在一定程度上丰富了"二十四间"的原有功能空间，也充分利用了废弃的老建筑，在保护的基础上再利用，为建筑创造了新的价值。

"二十四间"改造计划 | Renovation Plan for Twenty-four Houses

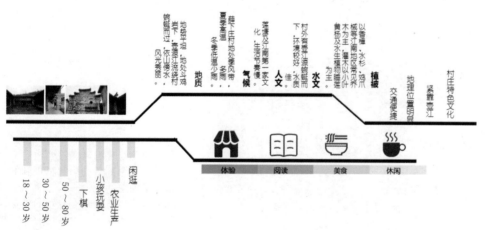

薛下庄村当前的主要人群以50～80岁的老年人为主，根据考察后对人群活动的了解，结合当地气候、水文、地质、人文、植被等特征，希望通过赋予体验、阅读、美食和休闲的功能来达到既满足原住民需求，又能提升当地旅游经济的目标。

室内分析 | Interior Analysis

平面功能分区 | Plane Function Division

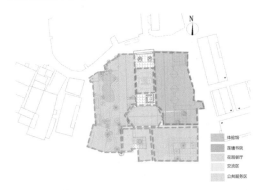

体验馆家具分析 | Experience Furniture Analysis

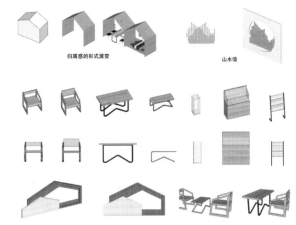

屋面形式演变家具

体验馆一层分析 | Analysis of Experience Floor 1

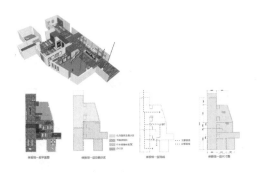

体验馆剖面 | Sectional Drawing of Experience

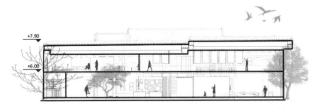

 体验馆一层以村庄历史文化展示、浦江书画、竹木根雕体验为主，二层以茶文化展示体验为主。

体验馆二层分析 | Analysis of Experience Floor 2

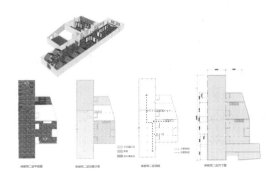

体验馆效果图 | Rendering of Experience

餐厅轴侧 | Side Axis of Restaurant

餐厅范围

餐厅中设置了六人、八人、十人的包厢，满足不同人群的需求，一层中心区域设置了一块抬高空间，丰富了空间形式，保证了顾客的私密性需求。

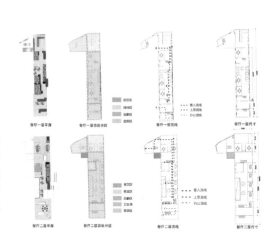

餐厅剖面 | Sectional Drawing of Restaurant

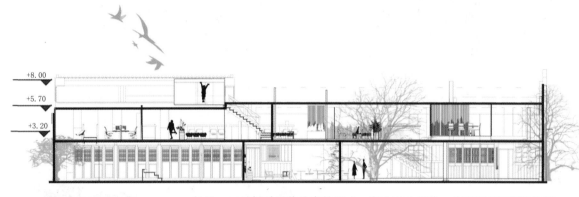

餐厅一二层都采用落地窗的形式，充分引入光线，也可以将西侧花园的景色更好地展现在顾客眼前。

餐厅效果图 | Rendering of Restaurant

莲塘书院分析 | Analysis of Liantang Academy

书院功能流线分析
| Analysis of Function Flow Line of Academy

书院剖面 Sectional Drawing of Academy

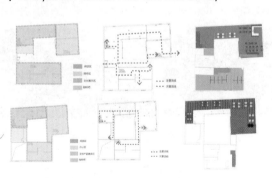

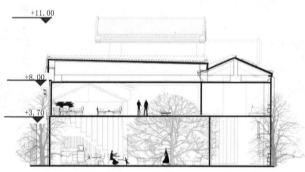

书院效果图 | Rendering of Academy

导师点评

王琛 Wang Chen

本案是新时期多样的社会背景下，对保护与活化利用乡村物质和文化遗产途径的积极探讨。方案深入挖掘和系统梳理了场地特质与文脉特点，合理推敲了置换功能，特别关注到活态的民众日常生活，自然地化作方案立意"合游"之中，使寿终正寝的老建筑在保护和再利用后得以与当下乡村再次"合体"并焕发出多元价值。设计协调了新旧空间与使用方式的协调转化关系，兼顾了建筑及其内外环境整体，这既复合该类项目的实际特点，也体现作者思考的全面性与专业综合能力。成果表达中，"化整为零"的表述之外欠缺对环境整体的清晰交代，也存在些许相关联的图纸前后不一等情况。这一方面使包括整体空间流线等问题存在与否尚未验证，另外一方面或可导致读者对方案本身的理解较难"化零为整"。

This project represents a positive exploration on the conservation, activation and utilization of rural materials and cultural heritages under the social background of diversity in the new era. The design deeply excavates and systematically hackles the characteristics and cultural features of the site, rationally deliberates the replacement function, and pays special attention to the people's positive daily life. Naturally, the above is "concretized" in the project, once again combining the rural buildings to be demolished together through their conservation and reuse, giving them multiple values. The design coordinates the transforming relationship between the old and new spaces and mode of occupation, and gives equal consideration to the buildings and the indoor and outdoor environments. Not only is this in line with the actual features of such projects, but this reflects the comprehensiveness of the designer's thoughts and professional comprehensive ability. In the statement of results, "the whole is broken up into parts", but the entire environment isn't clearly described, and some parts of the drawing are inconsistent with one another. On the one hand, the existence or non-existence of the overall spatial streamline has yet to be verified. On the other hand, readers may have difficulties in "gathering parts into a whole" while reading the systematization.

专家点评

王炜民 Wang Weimin

浙江工业大学第一组同学完成的《合游——"二十四间"老建筑保护与再利用设计研究》的设计成果围绕老建筑保护与再利用这一主题，关注乡村振兴发展，结合浙江民居的文化传统保护，与发展乡村旅游产业紧密结合，提出发展乡村旅游经济新业态，着重在利用薛下庄村现有空置的民居资源与空间环境，打造老建筑的保护与更新再利用相结合的民居改造设计方案。

设计小组同学采用了以推动乡村发展为导向的设计理念，对设计基地进行了深入细致调研工作。能够针对新时期乡村振兴发展建设中，如何发挥老建筑应有的作用进行了有益的探索，提出将"二十四间"原有的居住功能改造为发展乡村旅游产业为主题的展示、交流、茶饮、小型会议、接待等空间环境，使乡村闲置废弃的老建筑得以保护和赋予新的使用功能。他们基于这样的需求关系对老房子进行了重新审视和新设计的思考，建筑环境改造设计突出"合游"的主题与思路，提出了老房子改造中建筑与新功能、建筑与行为、环境与体验等方面设计构想与方案，以及探讨了如何在建筑环境改造中得以实施。设计方案通过调研、问题分析、方案表达，设计成果具有较为清晰和完整的表现力。

Zhejiang University of Technology Team 1's project is titled Heyou—Conservation and Reuse Design of "Twenty-four" Old Houses, which is about the conservation and reuse of old buildings. With focus on rural revitalization, the group puts forward an idea to develop the rural tourism economy by combining the protection of Zhejiang residents' cultural traditions with the development of the rural tourist industry. The group makes a residential building renovation scheme for the conservation and reuse of old buildings based on the idle dwelling resources and spatial environment in Xuexia Village.

The students in this group make in-depth research on the site under the design idea of promoting rural development. They make a helpful exploration on how to give full play to the functions of old buildings in the new era, when the countryside is revitalized, and propose transforming the "twenty-four" houses into rural tourist spaces for display, communication, tea tasting, small-sized meeting holding and reception, so that the idle rural old buildings could be protected and given new features. Based on these needs, they re-examine and think about the old buildings in a new way, and make an architectural environmental renovation design to highlight the theme and idea of "tourism". Then, they put forward a design idea concerning buildings and new features, buildings and behaviors and environment and experience in the old houses, and discuss how to put the idea into practice in the course of architectural environmental renovation. The design is based on research, problem analysis and project expression, and the design results are clearly and completely expressive.

同济大学图书馆室内外环境更新改造设计
Conservation and Regeneration Design of the Indoor and Outdoor Environment of Tongji University Library

高　　校：同济大学
College: Tongji University

学　　生：张奕晨、毛燕、潘蕾、Dana
Students: Zhang Yichen, Mao Yan, Pan lei, Dana

指导教师：左琰、林怡
Instructors: Zuo Yan, Lin Yi

课题评价：优秀
Achievement: Excellent

学生感悟
Student's Thought

张奕晨

感谢在同济大学的这几年，给了我充实而又丰富的校园生活。同时最后一个学期的毕业设计也让我收获了很多知识以及经验，也感谢在这过程中不断帮助我的同学和老师们。

毛燕

感谢毕业设计这一学期的经历，让我更充实和丰富。

潘蕾

为期三个月的毕业设计让我受益匪浅。感谢我的指导老师三个多月的悉心指导和帮助，感谢我的组员的陪伴与扶持，也感谢同一个课题的助教和成员们的帮助和关心。

Dana

非常感谢我能参与到这次的毕业设计项目中。同时，我也想要感谢我所有的老师和组员，非常开心能够和她们一起学习，一起进步。

总体规划 | Overall Planning

- ○ 室内家具布局改动
- ● 结构部件拆除
- ● 室内空间形式改动
- ○ 非结构部件拆除
- ○ 改建

拆：大厅连廊、屋顶
　　书库部分楼板、部分书搁
　　板、光导管
留：所有结构体系
　　书库钢书架体系
改：大厅连廊形式
　　书库功能、屋顶采光口
　　所有空间布局
建：大厅屋顶

私密　　　　　　　　公共

私　密：封闭研习室
　　　　影音室
　　　　朗读间
半公共：部分围合的阅览
　　　　空间
　　　　休息空间
公　共：交通空间
　　　　合度低的讨论空间
　　　　交流、休息空间

　　天窗
　　　玻璃顶
　　　　　光井
● 点状人工光
● 带状人工光
● 局部照明

自然光：大厅、玻璃廊通过玻璃顶引
　　　　入自然光
　　　　书库通过天窗、采光井引入
　　　　自然光
人工光：依据空间布局形式采取点
　　　　状、带状灯具
　　　　依据具体功能需求，局部增
　　　　加照明

历史

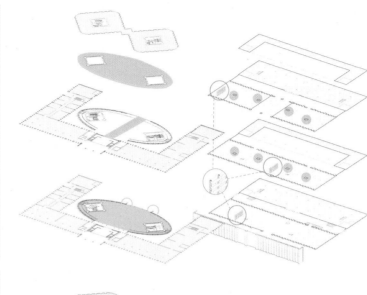

尺度

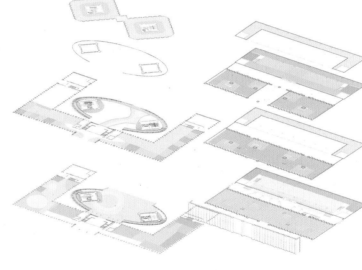

光环境

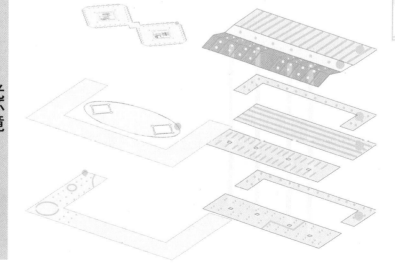

國粹賡續卷——歷史建築保護與再利用設計

答辯展示

大厅设计 | Lobby Design

大厅为图书馆中较为活跃与特殊的区域,其作为交通核心和服务中心的地位使得其在设计上有别于一般阅览空间的设计。

本次设计将在大厅中赋予更多学生活动的场所,增强其活跃性与开放性,提倡人与人面对面交流,为其创造积极活跃的空间。

设计策略 | Design Strategies

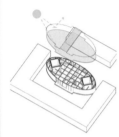

总体策略: 以保留和改建为主,部分拆除原大厅屋顶和二层连廊。

材质表: 以灰色白色木色材质为主,辅以亮黄色。

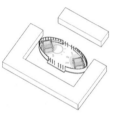

空间布局: 围绕两个核心筒做空间划分,并加入中央舞台。

私密度: 公共活跃度以舞台为核心向外扩散,围绕核心筒有些许较为私密的空间。

光照策略: 以玻璃顶取代原实心屋顶,充分引入顶部天光。

节点设计分析 | Node Design and Analysis

舞台模式　展览模式　书吧模式

舞台采用了多功能灵活性设计,设计了不同场景下的多种平面布置。

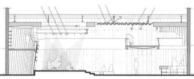

光环境设计 | Lighting Environment Design

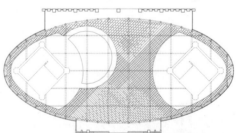

根据内部空间活动以及太阳轨迹研究,在大厅的玻璃顶下设计了具有遮阳功能的遮阳百叶吊顶。

其可以遮蔽夏季上午直射光,下午全年西晒直射光,而引入春秋冬季上午的部分直射光,调节室内光环境。

书库 & 玻璃廊 & 科技阅览室设计
| Book Storehouse & Glass Corridor & Science Reading Room Design

设计策略 | Design Strategies

书库
- 藏书 → 阅览 / 展览
- 结合书架改善功能

玻璃廊
- 各种功能融合
- 暴露书库清水砖墙

科技阅览室
- 丰富阅览模式

光环境设计 | Lighting Environment Design

科技阅览室—人工照明

1F 天花平面图

1F 局部照明平面图

灯具选用

改造前　改造后　t5 LED 灯管
表面连续凹凸纹的亚克力灯罩

管道状灯具
消防管道
风管
夹层点状灯具
桌面点状台灯
桌面带状台灯
落地灯
吊灯

整体 & 局部照明
发光 & 露明管道

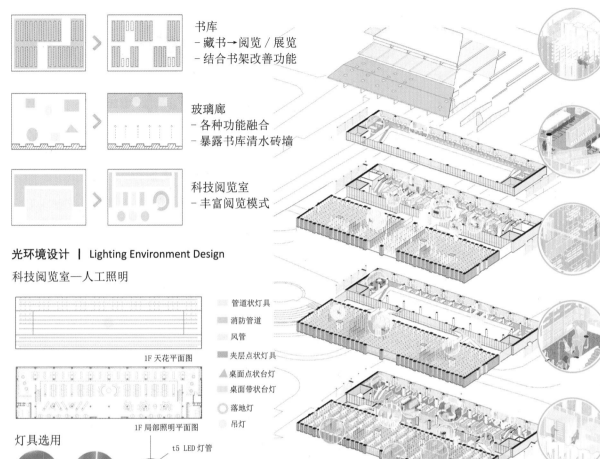

书库——自然光 & 人工照明

原状　拆局部楼板　置入亚克力柱　包裹玻璃

阅览座位
拆部分书搁板，
加座垫

原状

展柜内板
拆部分书搁板

展柜
玻璃包裹，
内置光源

采光井
亚克力柱
连接钢丝

展柜、座椅 & 天窗布置
展柜布置图
座椅与大窗布置图

天窗 & 座椅

书架灯光系统
T5 LED 灯管
延长原搁板
T5 LED 灯管

展柜灯光系统

缓解空间压抑感；
提高空间整体性；
人工光稳定柔和；
自然光增添趣味。

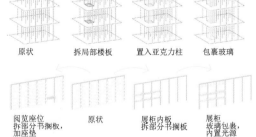
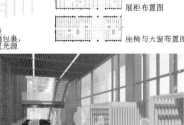

裙房设计 | Skirt Building Design

设计策略 | Design Strategies

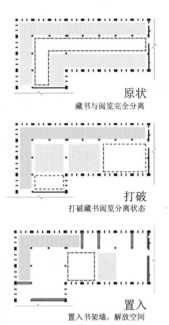

原状
藏书与阅览完全分离

打破
打破藏书阅览分离状态

置入
置入书架墙，解放空间

丰富
丰富阅览模式

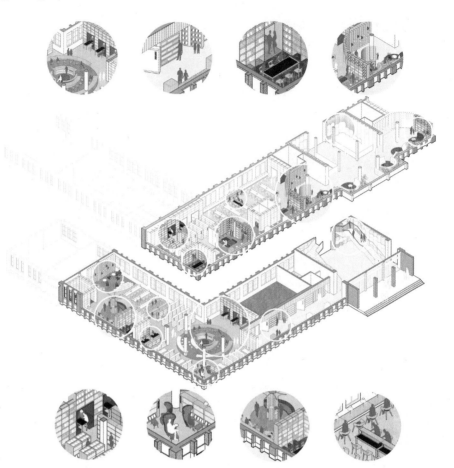

功能布局分析 | Functional Layout Analysis

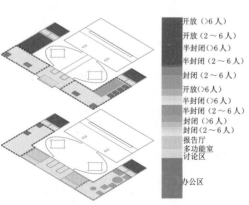

开放（>6人）
开放（2~6人）
半封闭（>6人）
半封闭（2~6人）
封闭（2~6人）
开放（>6人）
半封闭（>6人）
半封闭（2~6人）
封闭（>6人）
封闭（2~6人）
报告厅
多功能室
讨论区
办公区

光环境设计 | Lighting Environment Design

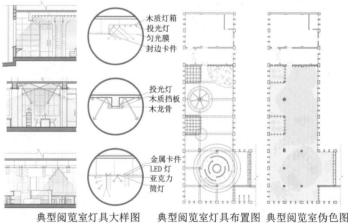

木质灯箱
投光灯
匀光膜
封边卡件

投光灯
木质挡板
木龙骨

金属卡件
LED灯
亚克力
筒灯

典型阅览室灯具大样图　典型阅览室灯具布置图　典型阅览室伪色图

塔楼设计 | Tower Design

塔楼位于整个图书馆的制高点，于 80 年代加建，整体悬挑超过了 8 米，获得了当时的多项创新结构奖项。

其内部作为阅览空间，我们对此进行了整体的改造。希望能改变塔楼单一的阅览模式，增加深阅读空间以及舒适的休闲空间。

设计策略 | Design Strategies

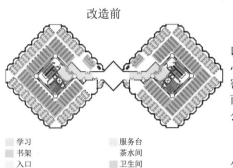

改造前
- 学习
- 书架
- 入口
- 服务台
- 茶水间
- 卫生间

改造后
- 开放阅览
- 私密阅览
- 舒适阅览
- 入口
- 茶水间
- 书架

平面布置以核心筒为核心布置较为私密的阅览空间，而靠窗则布置公共座位。

改造后减少了书架的数量和面积，但是增加了大量的舒适性座位和较为私密的深阅读座位。

同时增加了多样化的阅览模式，和多样化的学习休息空间。

家具设计

天花平面图

灯具布置图

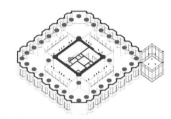

家具布置图

书架布置图

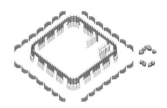

结构墙体图

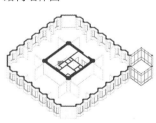

平面布局图 | Plane Layout

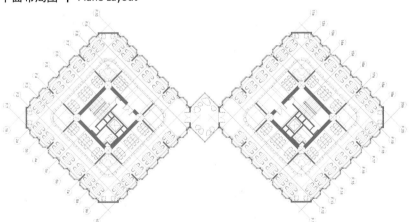

平面图 | Floor Plan

夹层平面图

二层平面图

夹层平面图

一层平面图

导师点评

吕勤智
Lv Qinzhi

同济大学第二组同学完成的《书+——同济大学图书馆室内外环境改造》的设计成果围绕"书"这一主题，以问题为导向，进行了深入细致地调研工作，对传统大学图书馆的功能与当代大学生对图书馆需求关系进行了重新审视和新的思考，能够针对传统理念上的图书馆如何在新时代发挥特有的作用进行了有益的探索。老图书馆建筑环境改造设计突出"书+"的主题与思路，以此拓展和探讨了新时代图书与知识、图书与学习的关系，研究了图书馆环境改造中书与行为、书与心理、书与体验、书与感受等图书馆新功能的作用发挥等相关命题如何在建筑环境改造中得以实施，方案设计的出发点和切入点正确；设计方案按照提出问题、分析问题、解决问题的方法推进设计研究的深入与过程展开，步骤和方法具有逻辑性和可操作性；设计方案通过调研与数据收集、问题比较与分析、设计理念确立与拓展、设计方案推敲与提出、方案图纸与模型表达等设计环节的推演，使成果的呈现具有清晰的表现力，表现出设计团队同学们扎实的专业基本功、良好的协同工作能力、建筑设计的创新能力，以及设计成果的表达能力。

Tongji University's Team 2's Books+——Indoor and Outdoor Environment Renovation of Tongji University Library keeps to the point of "books" and makes an in-depth research around problems. The designer re-examines and rethinks about the traditional university library's function and contemporary students' needs for the library, and then carries out a helpful exploration on how to give the conventional library special features in the new era. The renovation design of the building environment of the old library highlights the idea of "books+", by which the designer expands and explores the relationship between books and knowledge, as well as between books and study, in the new era, and researches how to give play to the new functions of library in the renovation of the building environment, including the relationship between books and behavior, books and mentality, books and experience, and books and feelings. The design has a proper objective and idea; the designer carries out research from the shallower to the deeper by raising problems first, then analyzing the problems and finally solving the problems. The steps and methods have logicality and operability; the results are clearly expressive, due to research and data acquisition, problem comparison and analysis, design idea establishment, design scheme deliberation, drawing making and model formulation. This reveals the designer's solid specialty foundation, good collaborative capacity, innovation ability in architectural design, and ability of expressing design results.

专家点评

李正涛
Li Zhengtao

同济二组的题目是同济大学图书馆室内外环境改造，设计的主题为"书+"。发现问题、思考并研究问题、提出解决方案，设计通过这个思路展现出来。思考过程严密，成果详尽完善，是一个优秀的毕业设计。

整个设计过程逻辑清晰，从使用者角度发现图书馆室内外环境的实际问题，如阅览空间品质低、纸媒阅览衰落等，问题真实且具有代表性；设计全面而认真的分析了问题，倡导图书馆以纸媒阅览为主，电子为辅，增强阅览的多样性，最后提出的设计方案也非常细致，传承本馆历史，提出改造策略，并对每个主要区域、各个专项做了相应的设计，呼应主题，力图架构纸媒与数字化的关系，架构人与人、人与书的关系，从而使图书馆能成为书之家，人之家。

Tongji Team 2's project is indoor and outdoor environmental renovation of Tongji University Library. The design theme is "books+". The design unfolds by identifying problems, thinking about and researching the problems, and proposing solutions. It is an excellent graduation design considering that the thinking process is precise and the results are detailed.

In the whole design process is a clear logic. Practical problems in the indoor and outdoor environment of the library, such as the low quality of reading space and the decline of paper reading, are identified from users' perspective. So, the problems are real and typical. The designer analyzes the problems seriously in a comprehensive way, suggests using print books as the main reading materials, while e-books as auxiliary materials, and finally proposing a detailed design scheme. To inherit the history of the palace, the designer advances a renovation strategy and makes a design for every main area and each item to echo with the theme, in an attempt to build a relationship between print books and digitization, between people, and between people and books, so that the library could become a home of books and a home for people.

30个模块+7个技术——云县传统民居更新设计

30 Modules+7 Technologies——Renewal Design of the Traditional Dwellings in Yun County

高　　校：华南理工大学
College:　　South China University of Technology
学　　生：王铭、张豪元、王璐瑶
Students:　　Wang Ming, Zhang Haoyuan, Wang Luyao
指导教师：石拓
Instructors: Shi Tuo
课题评价：良好
Achievement: Good

学生感悟
Student's Thought

王铭

　　我们在这次设计中不仅仅是为了做一个设计而已，而是希望我们所给出的这种思路能为今后农村城镇化发展提供一种设计导则。

张豪元

　　从去云南实地调研再回来，我们进行了大量的分析例举不同的空间可能性。为村民们规划出不同可能性的功能区域应该是这次设计最复杂的一部分，因为我们需要知道他们实际的需求以及为了保留他们当地的传统特色文化，我们不能去毁坏这些历史文化。

王璐瑶

　　通过这次设计，我才真正意识到作为设计师我们应该站在使用者的角度上去为他们做真正需要的设计。

昔宜村背景介绍
Background Information of Xiyi Village

本案的中国村庄位于云南省临沧市云县昔宜村，村落临近澜沧江，村民多以渔业和茶叶为经济收入，全村三千余人，占地四十多平方公里。是漫湾电站库区移民后期扶持的重点村。古村原状采用乡土材料建设，充满地域特色，周边旅游资源与生态景观资源丰富，加上其交通便利性，国家政策扶持，未来的旅游业发展前景不可限量。在全国精准扶贫，脱贫攻坚的工作开展之下，乡村传统民居有政府指导的新居方案，但现在村民的生活质量提升，又有旅游方面的需求。政府指导的新居方案不但不适应农民生活，而且村民自己新建的新居失去古村特色。

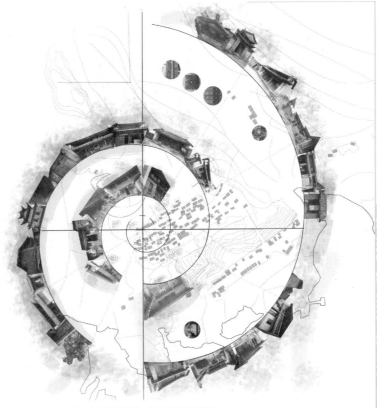

实地踏勘调研分析 | 具有优秀文化遗产
Field Research and Analysis

提出问题 | 新建筑问题与传统建筑问题
Question Raising

建筑类型丰富

三合院式　四合院式

建筑文化丰富

在墙上的鸡窝

石制水缸

火塘

叉叉锅

建筑装饰丰富

新建筑问题

· **千村一面**

目前的统一规划方案中未能充分考虑当地地域特色，同时追求整齐划一的效果，因此造成昔宜村与全国其他村庄一样千村一面的局面。

目前的临沧市云县漫湾镇昔宜村委老村组省级示范村村庄规划样图

· **功能单一**

一层平面　　二层平面

面对每一个不同的家庭，对功能的需求也不同。有些家庭的功能倾向自住，有些可能倾向民宿、餐饮等其他的功能，但在目前统一规划方案中的平面都是一样的。

临沧市云县漫湾镇昔宜村委老村组省级示范村村庄规划样图平面图

· **现代材料失去传统风貌**

在政府的统一规划方案中，对新建筑屋顶与墙体的材料与做法均有限制要求，现代建筑材料在建筑外立面的简单运用使村落失去传统建筑风貌。

建筑装饰丰富

- ⚡ 防雷击
- 🧯 消防安全
- 🌡 保温隔热
- 🔊 隔音效果差

卫生与污水处理问题
传统建筑由于建筑年代与当时生活水平原因，基本没有独立厕所等配套，大小便主要在猪圈等牲畜棚里解决，洗澡则是男女错开在澜沧江边进行。

防虫防腐问题
由于建筑年代已久，各种设施构建不同程度腐烂，加上室内地板多为已破损的水泥板，容易隐藏虫害。

安全抗震问题
由于建筑年代已久，传统建筑多为木构，各种设施构建不同程度腐烂，在恶略天气情况下居住易发生漏雨、松动等危险。

设计策略 | 千村一面如何留住乡愁
Design Strategies

村民访谈

能否像拼积木一样的功能模块化给村民改造建议？

可能村民更多需要的是一个设计导则

而不是具体的某一个建筑

改善目标

·改善居民生活
即在保留原来建筑外观的前提下，通过改善居民居住环境增加新功能。

·传统文化传承
把旧建筑中传承很久的文化功能提取出来，在改造过程中不能遗失掉。

·满足旅游发展需求
满足由旅游业发展带来的大量城市游客的需求。

即村民根据自己的实际情况选择自己的房子的功能组合方式

以昔宜村罗良海宅为模板分析 模块化优势分析

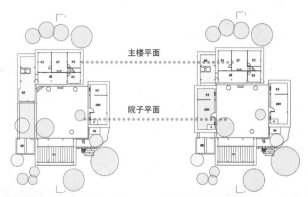

昔宜村罗良海宅现状一层平面　　昔宜村罗良海宅原貌一层平面

模块化可行性分析

云县传统民居使用木构架承重，墙体一般仅用作维护，并不承受屋顶重量。山墙和木构架是两套体系，因此可以见到墙倒屋不塌，或是墙塌了而木构架不倒的现象。

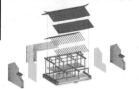

·框架结构
·墙体可拆
·灵活性强

·可以保留传统风貌
改变千村一面局面与解决材料问题。

·功能可选
解决功能单一性的问题，提供更多选择的方式。

·成为建造设计导则
按需的简单易行，灵活的功能搭配。

模块空间尺寸演变 | 以昔宜村罗良海宅四合院主楼为模板分析
Dimensional Change of Module Spaces

2600mm×3200mm　　3200mm×3200mm　　5800mm×3200mm　　8400mm×3200mm

模块策略举例 | 5大主功能模块 + 多个辅助功能模块
Examples of Module Strategies

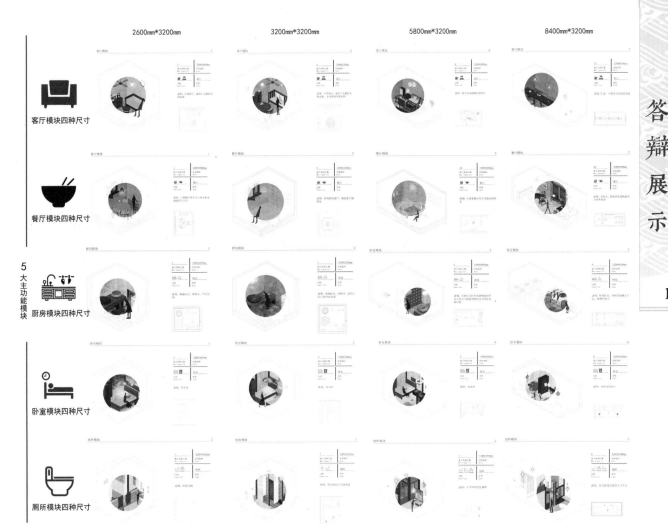

多个辅助功能模块：超市模块、儿童间模块、酒吧模块、展厅模块、公共洗衣房模块、卡拉OK模块、养殖模块

模块组合举例 | 村民自住模块组合举例 + 民宿等其他功能模块组合举例
Examples of Module Combination

组合原则

人数问题
改造后的房屋要住进多少人？

房屋定位/改造目标
改造房屋未来主要目标是盈利还是住的更舒服还是两者都有？

噪声污染
小酒吧、餐厅这些产生噪声的功能空间要远离卧室。

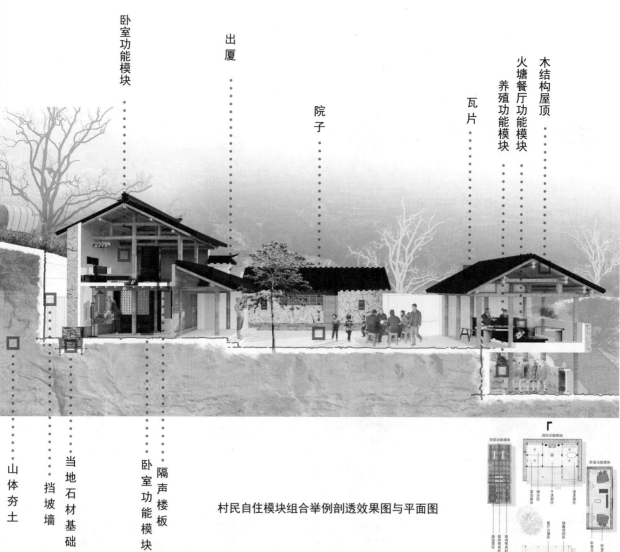

卧室功能模块 · 出厦 · 院子 · 瓦片 · 木结构屋顶 · 火塘餐厅功能模块 · 养殖功能模块

山体夯土 · 挡坡墙 · 当地石材基础 · 卧室功能模块 · 隔声楼板

村民自住模块组合举例剖透效果图与平面图

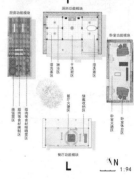

客厅功能模块

澡堂功能模块 卧室功能模块

民宿等其他功能模块组合举例剖透效果图与平面图

适应性改造 | 多个安全改造技术举例
Adaptive Renovation

卧室功能模块　厕所功能模块

客厅功能模块

泡澡池　淋浴区　卧室　干湿洗漱区　储藏收纳区　餐厅火塘区　棋牌娱乐区　客厅娱乐区

火塘民俗餐厅功能模块

选用耐力壁加固房屋结构，一个完整平面开间的四个角落支撑柱的两侧，加固位置如图

N 1:94

抗震

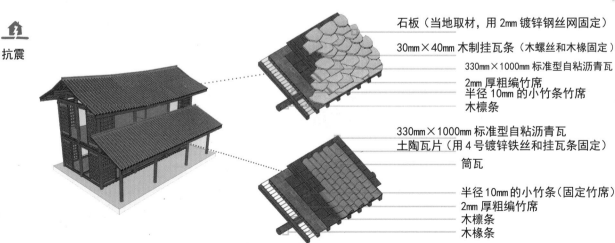

石板（当地取材，用2mm镀锌钢丝网固定）
30mm×40mm 木制挂瓦条（木螺丝和木橼固定）
330mm×1000mm 标准型自粘沥青瓦
2mm 厚粗编竹席
半径10mm的小竹条竹席
木檩条

330mm×1000mm 标准型自粘沥青瓦
土陶瓦片（用4号镀锌铁丝和挂瓦条固定）
筒瓦

半径10mm的小竹条（固定竹席）
2mm 厚粗编竹席
木檩条
木橼条

國粹賡續卷——歷史建築保護與再利用設計

答辯展示

防虫防腐

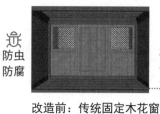
改造前：传统固定木花窗

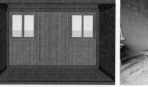
改造后：推拉玻璃窗

窗户的改造，将固定木窗改造为平开玻璃窗，或增加推拉玻璃窗。

桐油作隔潮防腐剂，添加2%五氯酚钠或菊酯（《古建筑木结构维护与加固技术规范》）。

墙体保护

- 墙体
- 水泥：河沙：黄土：石灰（1：4：5：1.3）混合砂浆打底
- 纤维网
- 水泥：河沙：石灰（1：7：0.8）砂浆
- 墙漆

室内外地铺板装

- 30mm厚石板 400mm×400mm
- 40mm厚干硬性水泥砂浆
- 素土夯实

- 间填水泥砂浆
- 石头
- 素土夯实

- 18mm厚木板
- 20mmXPS
- 2mm粗编竹席
- 半径10mm的小竹条（固定竹席）
- 木檩条

- 20mm厚石板 400mm×400mm
- 30mm厚干硬性水泥砂浆
- 素土夯实

隔声

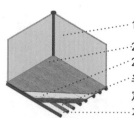

- 18mm厚木板
- 半径10mm小竹条（固定竹席）
- 2mm粗编竹席
- 木檩条
- 20mmXPS

隔声板设置位置：
地板——一层与二层间楼地面——XPS挤塑聚苯板
墙体——房间隔墙处——双层墙体

污水处理节点分析

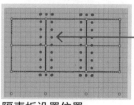

- 污水进水口
- 粪渣口
- 出水口
- 美人蕉
- 黄花蔺
- 风车草
- 1350mm
- 2700mm
- 鹅卵石
- 化粪池（地下）
- 第一格
- 第二格
- 第三格
- 人工湿地（底部坡度1.3%）

导师点评

马辉
Ma Hui

华南理工大学同学针对云县传统民居能够以问题作为出发点进行设计思考，并总结出主要问题千村一面与功能缺失。同时针对问题建立设计逻辑，提出了有效的设计策略，即模块化的解决方式。作品中较好的表现了这一设计策略，从传统民居的各个功能空间出发，设计了相对适应当代村民生活的有效模块，体现了该组同学较强的设计能力与设计素养。

在更新改造方面，该组同学也提出了适应时代发展的改造技术，同时从材料、工艺、构造、环境友好几个方面具体的展示了这些技术的实现可能性，很好的保持了原始民居村落的传统风貌，同时又改善了村民的生活条件，提高了生活环境的质量。

然而，在群落的基础设施方面，只是在消防安全角度提出了消防水鹤等建设布局的改善措施。目前，乡村的居民生活污水由于洗涤剂大量使用，造成对村落土壤、河流、地下水等环境的严重污染与危害，此方案中针对村落居民的生活污水排放方面没有提出更好的有效应对策略稍显稀罕，希望在今后的学习与工作中能够更多的关注环境问题。

The students from South China University of Technology think about the design of Yun County traditional dwellings by starting with problems, and identify the main problems: similarity in appearance and absence of functionality. Moreover, they set up a design logic for the problems, and figure out a design strategy, namely a modular solution. This design strategy finds vivid expression in their work: they design effective modules suitable for the contemporary rural life based on various functional spaces in traditional dwellings, revealing their strong design capability and literacy.

In terms of renovation and transformation, the team advances opportune techniques, and shows the practicability of these techniques in several regards such as material, technology, structure and environment-friendliness, well retaining the traditional scene of the traditional dwellings, improving the villagers' living conditions, improving the quality of the living environment.

However, in terms of basic facilities, the designer just takes measures to improve the water cranes of fire from the perspective of fire safety. Currently, the rural environment, including soil, rivers and underground water, has been severely polluted by domestic sewage containing plenty of detergents. It is uncommon that this solution does not contain any effective countermeasure for domestic sewage disposal, so I hope that the designer will pay more attention to environmental issues in future study and work.

专家点评

姚领
Yao Ling

该组能够脚踏实地的针对乡村发展与改造中的建造碰到的具体问题做总结和分析，面对各方多种诉求和问题进行细致深入而系统的梳理，运用模块化和类型化的专业解决问题的思路值得肯定。

尤其是针对政府要求和村民的自我诉求之间的矛盾，具体的建造过程中碰到的主要和关键问题的技术化研究的思考和研究，都能够提供给项目积极而有价值的参考，能够很具体的技术化的落地解决乡村发展中的相关问题。

如果能够结合地区性特点，在其独特性和唯一性的价值基础上，在当地如何发展的设计策略中给出更有想象力和思考力的研究就更好了。

The team worked in a down-to-earth way, summarizing and analyzing the specific problems encountered during rural development and renovation, carefully and systematically reviewing multi-party demands and issues. It is worthy of approval that they finally used a modular and categorical idea to solve the problems.

Especially, for the contradiction between the government request and the villagers' demand, as well as the main and key problems encountered in the construction process, the group members think and research from a technological perspective, providing a positive and valuable reference for the project. They can solve the problems arising in rural development by specific techniques.

It will be good if an imaginative and cogitative study can be made on the design strategy concerning the local development based on the uniqueness according to the local characteristics.

历史建筑保护背景下建筑空间及其社区环境改造设计
Renovation Design of an Architectural Space and its Surroundings in the Context of Historic Building Conservation

高　　校：哈尔滨工业大学
College: Harbin Institue of Technology
学　　生：周子钦、周毓、赵斌、颜岩
Students: Zhou Ziqin, Zhou Yu, Zhao Bin, Yan Yan
指导教师：马辉、周立军、兆翚
Instructors: Ma Hui, Zhou Lijun, Zhao Hui
参赛成绩：良好
Achievement: Good

学生感悟
Student's Thought

周子钦

随着此次"室内设计6+"联合毕设的圆满收官，首先要感谢我们的指导老师马辉对我的悉心指导与帮助。其次，在设计过程中，我通过查阅大量有关资料，与同学交流经验和自学，并向老师请教等方式，使自己学到了不少知识，也培养了我独立工作的能力，相信会对今后的学习工作生活有非常重要的影响。

周毓

我非常高兴能参加这次"室内设计6+"联合毕设，一次经历，一次成长，这次比赛的整个准备以及参赛过程对我来说是一次美好的回忆。

很感谢这次"室内设计6+"联合毕设，让我有机会充分发挥所长，也认识到自己的缺陷和薄弱所在。感谢周老师和兆老师以及马老师对我们的悉心指导。在今后的学习道路上，我将继续捧着一颗热忱的心，努力做到更好！

赵斌

我认为，每一次经历都是生活给予的宝贵经验，是成长的必然。这次参加"室内设计6+"联合毕设也不例外。设计过程中，我在发现不足的同时，也借鉴了他人的经验，让我总结出四个字——越挫越勇。我还告诉自己：不要被自我感觉所蒙蔽，你还有许多要学习的地方，要多多努力，让自己变得越来越优秀。

颜岩

很荣幸可以参加这次"室内设计6+"联合毕设。首先我要感谢老师们的悉心指导。其次感谢和我一起参加竞赛的同学们，感谢他们在学习的时光中对我的帮助和鼓励。通过这次竞赛，让我知道我的设计还存在不足之处，也让我知道在设计这条路上我还有很远的路要走。不管设计这条路有多么的艰辛，我相信我会通过我的努力一直坚持下去。

概念生成 | Concept Generation

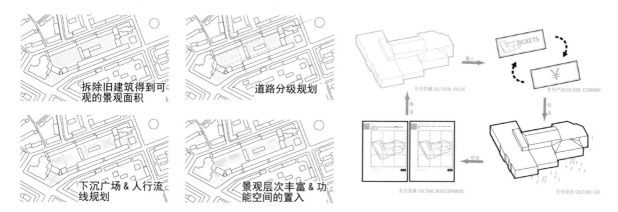

　　为了场地和文化宫地下相互产生联系，将场地进行下沉，从入口广场逐步下沉到3000mm的深度，北侧与文化宫地下通道相连，游客可以在欣赏完文化宫的剧目之后从退场通道进入到场地内部，进行游览和赏景。下沉广场也和社区活动中心相沟通，游客也可以直接参观社区活动中心一层的展览后进入下沉广场，之后进行景观上的游览。

总平面图 | General layout plan

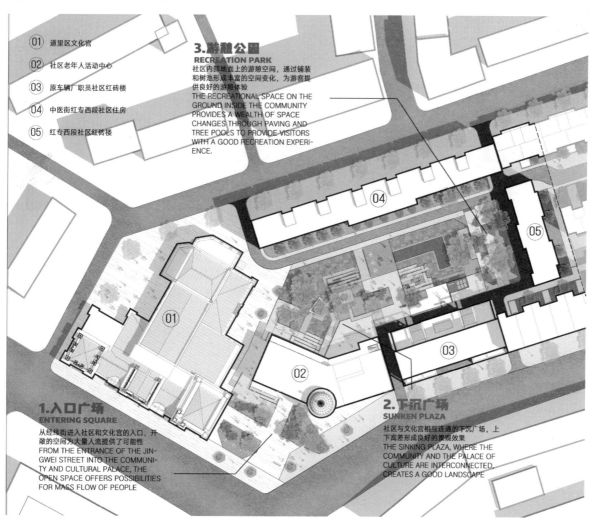

场地设计 | Site Design

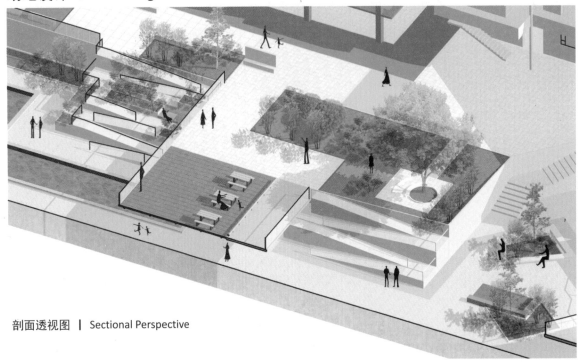

剖面透视图 | Sectional Perspective

　　游客可以从中医街、红霞街两侧进入广场。空间层次上增加了高差变化，丰富了景观的空间关系。植物上以林植和丛植为主，乔灌和地被植物相互结合呼应，营造层级关系，错落有致，丰富了植物种类间的层次，带来更好的观赏效果。

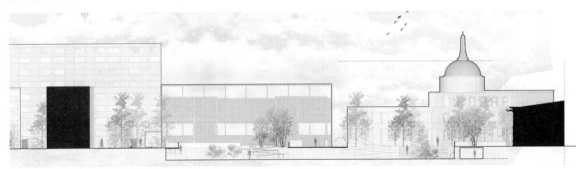

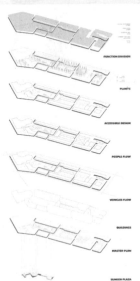

　　乔木以常青木和小叶乔木为主，考虑到寒地特征，主要观赏性在于观赏枝干，配以银杏树营造季节性变化。灌木以暴马丁香或紫丁香以及金银忍冬为主，花期在4月到6月，相对适合哈尔滨气候，可以在春季观花，夏季赏叶，秋季观果，冬季赏干。地被也多为4月到6月开花，以适应场地位于寒地的特征。

道里文化宫室内改造 | Interior Renovation of Daoli Cultural Palace

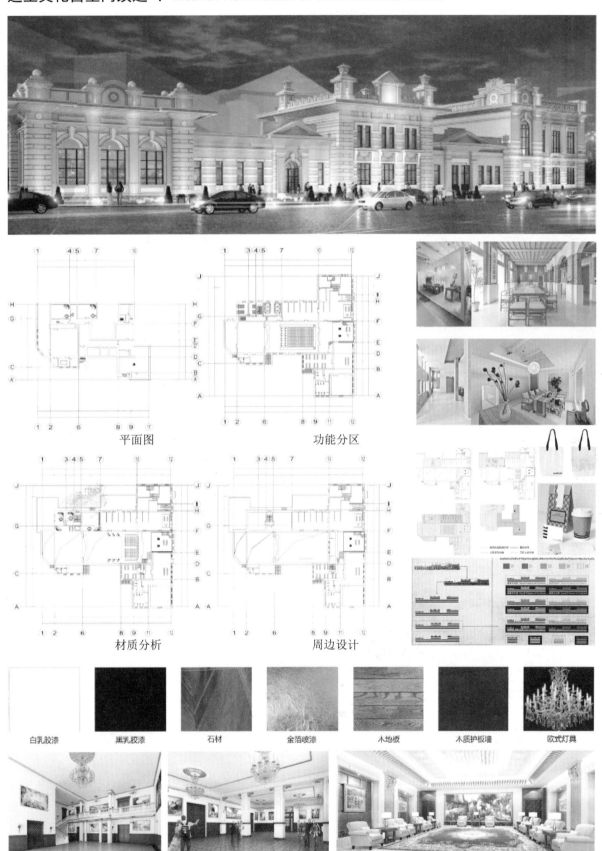

老年活动中心室内改造 | Interior Renovation of Senior Citizen Activity Center

功能分区

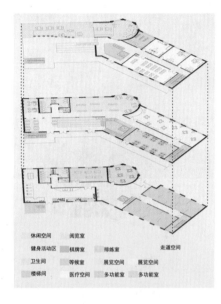

概念生成

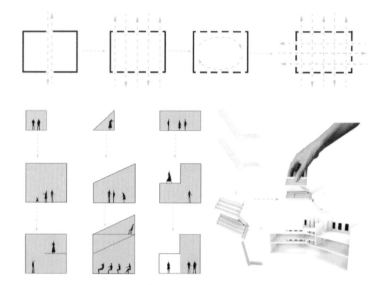

室内外休息室

室内活动室

展厅

儿童活动室

流线分析

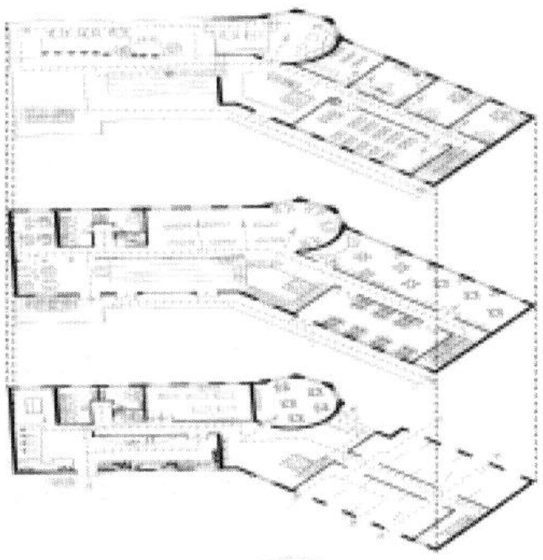

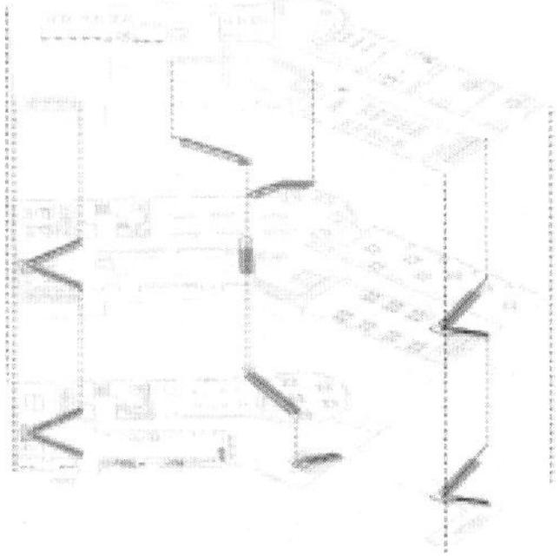

方案分析

卫生间下水设计中，考虑到传统的门槛设计会使坐轮椅的老人无法使用、高龄老人容易绊倒，所以采用了在地面留出一条排水缝。

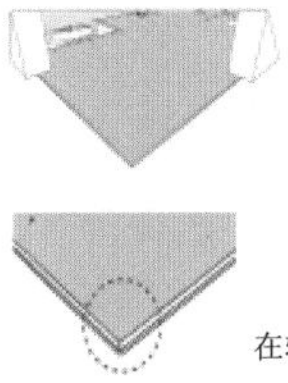

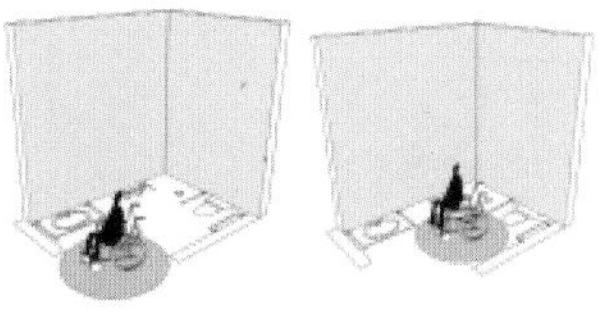

在轮椅回转半径的设计上，利用双拉门提供轮椅转半径的距离。

场地通行流线

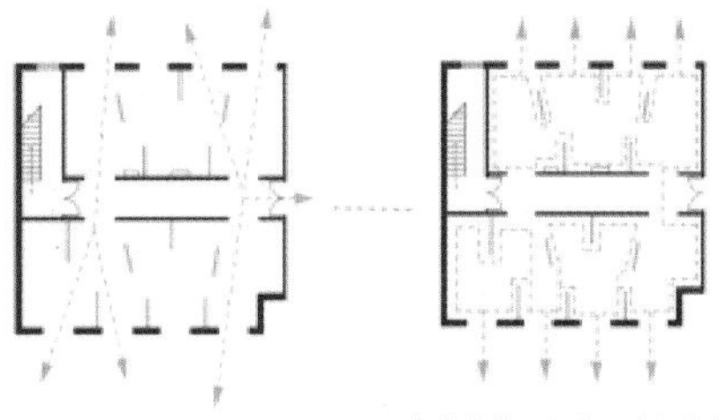

参观展览路线

考虑人体工程学，确定信息大小与安置位置。

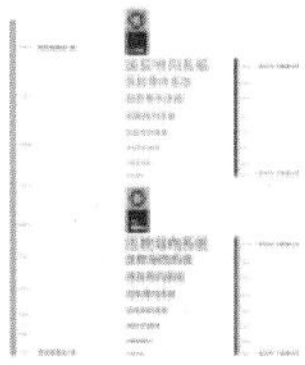

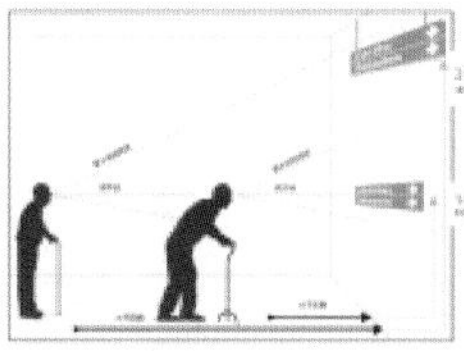

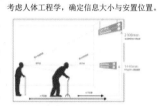

图示规格为 150mm ≥ a ≥ 30mm。

标识中文字规格为 100mm ≥ a ≥ 20mm。

确定符号系统视觉识别尺度标准。

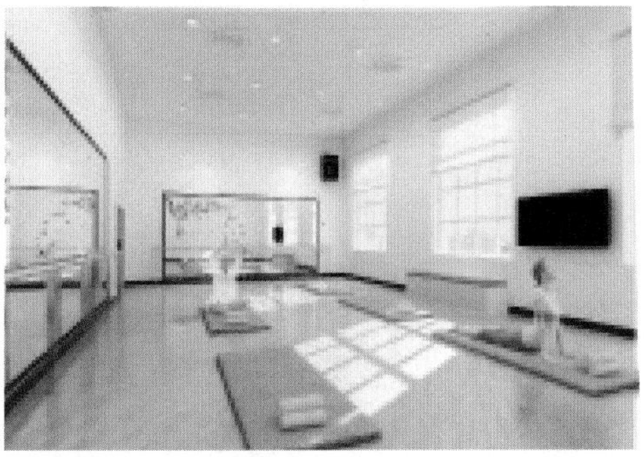

舞蹈排练室

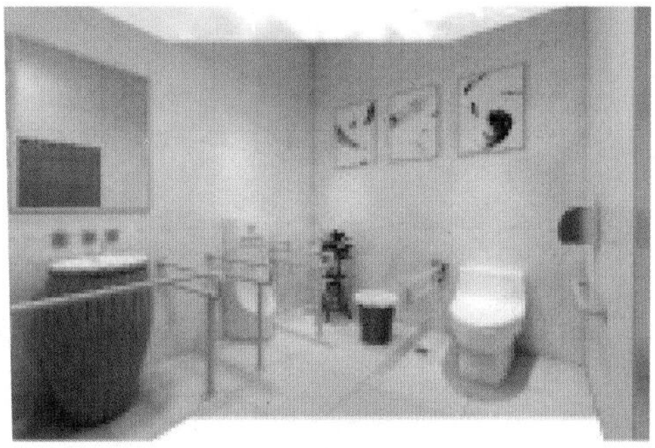

卫生间

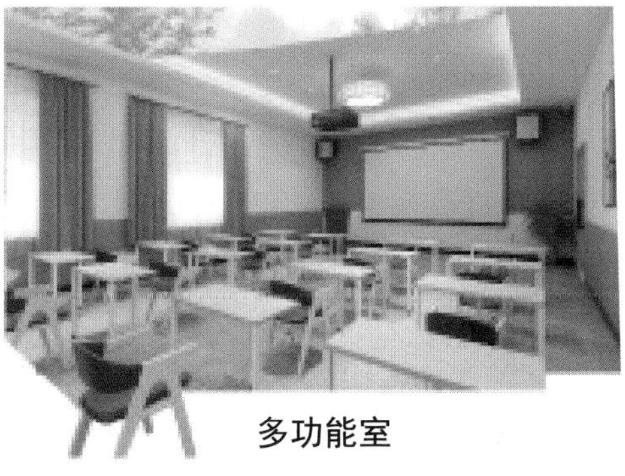

多功能室

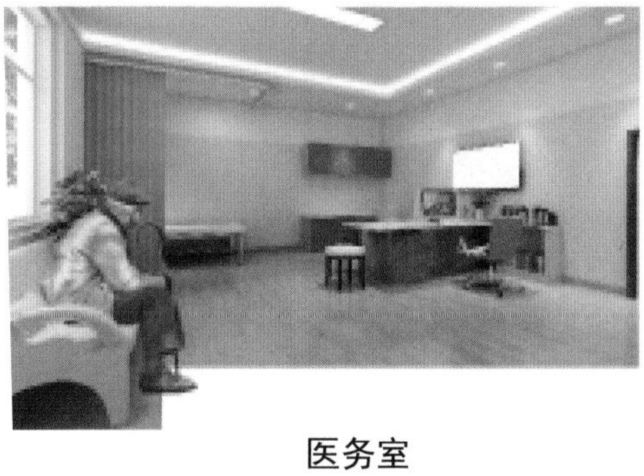

医务室

社区老年活动中心原始、现有平面对比

红砖楼改造 | Renovation of a Redbrick Building

功能分区 & 流线分析

红砖楼原始、现有平面对比

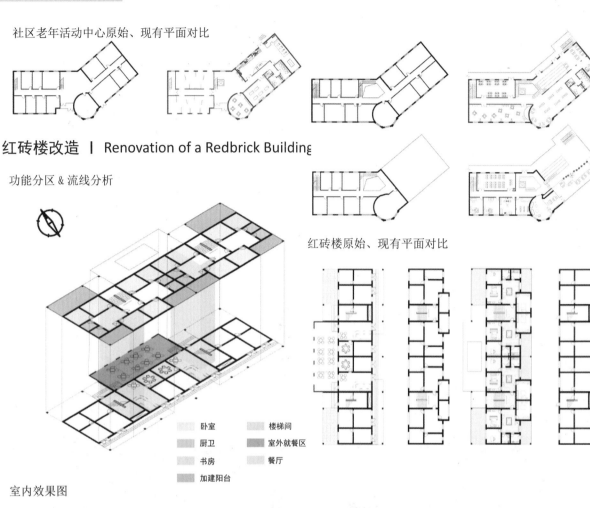

卧室　　楼梯间
厨卫　　室外就餐区
书房　　餐厅
加建阳台

室内效果图

卫生间效果图

厨房效果图

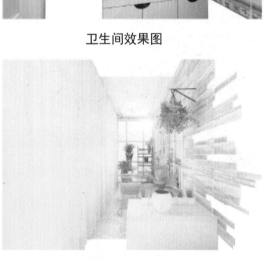

加建阳台效果图

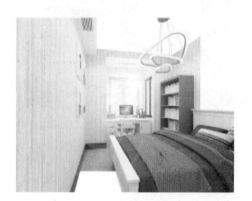

次卧效果图

导师点评

王敏
Wang Min

该组毕业设计作品以哈尔滨道里文化宫及其社区环境的保护性利用为题，设计思路清晰，空间分析完整深入，在不破坏原有历史建筑的前提下，巧妙地运用了"嵌套"的设计手法，赋予历史建筑以新的空间构成及功能，同时木质、金箔等材料的使用又保留了部分原有的建筑与文化特色，保护与开发共存。室内空间功能明确、流线合理，围绕目标群体有较突出的设计细部表现。设计方案系统完整，从建筑、景观、室内直至配套的周边产品设计，反映了其良好的综合设计能力及专业素养。

This team's graduation project concerns the protective utilization of Harbin Daoli Cultural Palace and its surrounding community environment. The designer thinks clearly, and makes a thorough in-depth spatial analysis of the building. On the premise of not destroying the historic building, the designer gives the historic building a new spatial constitution and function by using the nested design technique skillfully, and retains the original architectural and cultural characteristics by using wood and gold foils, achieving the goal of protection and development. The interior space has a definite function, a rational flow line, and outstanding details for the target group. The design scheme is systematic and complete, and the design of everything, including the building, landscape, interiors and peripherals, reflects the designer's excellent comprehensive design capacity and professional literacy.

专家点评

王野
Wang Ye

历史保护建筑的改造与再利用当下在很多城市都是迫切待解的问题，我想不应只是常规的整饬而是要赋予建筑新的生命，要充分理解该建筑的历史重要性及意义，让其成为城市历史的见证和记忆之外，同时也成为一种再生的城市建设发展新资源。

本案设计，同学们赋予了新的建筑生命，如芭蕾舞剧场、展厅、书吧、咖吧等很好的当代空间功能，用时也为城市提供了优质的市民活动空间和城市产业的更新空间。注意到有陈旧设备管线与系统的重新设计，保持了良好的建筑永续性。在原真性上需加强思考，如新空间的材质与工艺的选择、与原建筑风格是否有延续性与和谐性。在设计手法上要多思考，如在新空间上保留局部原建筑老墙的原态与新墙的对比效果以增加介入者的历史带入感。

Currently, the renovation and reuse of historic buildings is a problem that urgently needs to be addressed in many cities. In my opinion, we shouldn't just renovate an old building routinely, but should give it a new life, and make full sense of its historical importance and significance, to make it a witness and memory of the history of a city, and a regenerated new resource for urban construction and development.

In the project, the students give the building a new life, including good contemporary spatial functions such as ballet theatre, showroom, book bar and coffee bar. When used, the building also provides the city with a quality citizen space and a space for urban industrial renewal. Obsolete equipment, pipelines and systems are also resigned, maintaining architectural sustainability. Authenticity needs further thinking, such as the selection of materials and technology for new spaces, and whether the new architectural style is in line with the original one. The design technique needs much thinking. For instance, the original look of the old wall might as well be retained locally in contrast to the new wall, so that viewers could get a strong sense of history from the building.

韩城古城的保护与再利用设计
Conservation and Reuse Design of Hancheng Ancient City

高　　校：西安建筑科技大学
College: Xi'an University of Architecture and Technology
学　　生：曹玥玲、张景一、任晓贤、王玲子
Students: Cao Yueling, Zhang Jingyi
　　　　　Ren Xiaoxian, Wang Lingzi
指导教师：王敏、刘晓军
Instructors: Wang Min, Liu Xiaojun
课题评价：优秀
Achievement: Excellent

学生感悟
Student's Thought

曹玥玲

成功不是将来才有的，而是从决定做的那一刻起，持续积累而成。
　　　　　　　　　　　　　　　　　　——俞敏洪
所以感谢建大四年来对我的教务和培养，让我掌握了丰富的理论知识和制图经验，不断充实和完善自己。

张景一

大学是一副空白画卷，我用双手和勤奋描绘了属于自己的五彩青春。

任晓贤

在此次的设计中我收获到了很多，比如团结协作，对自己高要求。我相信，每次的努力，学到的东西都是受用一生的。

王玲子

所有跋涉，都是为了抵达。希望明天的你，会感激今天努力的自己。

建筑分类 | Building Classification

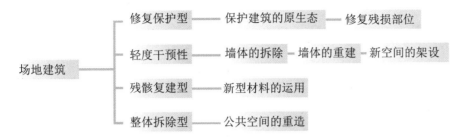

修复策略 | Renovation Strategies

保存较好建筑

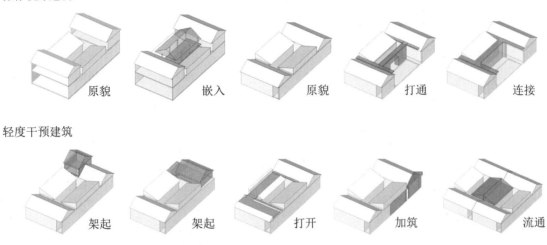

原貌　　嵌入　　原貌　　打通　　连接

轻度干预建筑

架起　　架起　　打开　　加筑　　流通

残损建筑

原貌　　材质对比　　材质对比　　防护　　架设

组合形式 | Combination Form

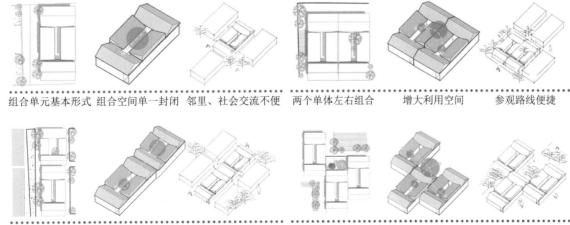

组合单元基本形式　组合空间单一封闭　邻里、社会交流不便　两个单体左右组合　增大利用空间　参观路线便捷

两个单体前后组合　拉伸纵向流线　增大活动范围　单个单体生成组团　空间层次丰富　连通建筑内外空间

印花袱子 | Printed Cloth Scarf

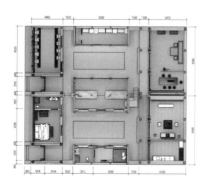

平面图

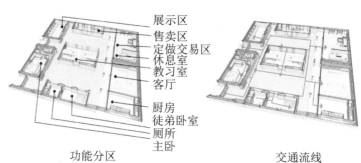

展示区
售卖区
定做交易区
休息室
教习室
客厅
厨房
徒弟卧室
厕所
主卧

功能分区　　　　　交通流线

印花袱子现状：
（1）韩城市是渭北"花袱子"生产最集中的地方。存活很多家庭作坊。
（2）自20世纪90年代末以后，花袱子手工印染生意在渭北开始逐渐减少。
（3）进入21世纪以来，可以说已濒临凋敝的危境。
（4）老艺人转行，盛世景象不复存在。

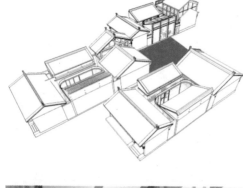

空间单元

社区服务中心 | Community Service Center

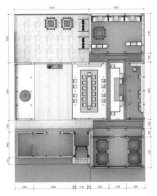

平面图

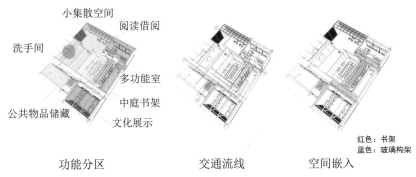

功能分区　　　　交通流线　　　　空间嵌入

红色：书架
蓝色：玻璃构架

社区服务中心在设计时，考虑到周边居民的需求，设置了小型的集散区域、读书室、藏书室，还为居委会提供了开会的多功能室等一系列功能区来满足周边居民的生活需求。

效果图

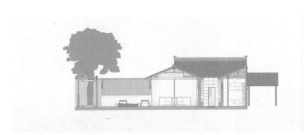

三连院子 | Three-space Courtyard

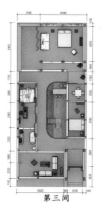

第一间　　　　　第二间　　　　　第三间

三联院落由三间院落和共享花园组成。

三连院子是由三间独立的四合院构成，三间院落围合构成了一块小空地，将此处设计为三间的共享花园，而三间院落的室内设计也围绕这块花园来设计。

第一间：在中庭部分架设了玻璃廊道，链接各个房门，既不影响通风采光，又可在冬季抵御寒冷。
第二间：对坍塌的屋顶进行修复与部分的抬升。
第三间：作为餐厅的东厢房全部采用全新的材料，属于半开敞空间。

墙体改造示意图

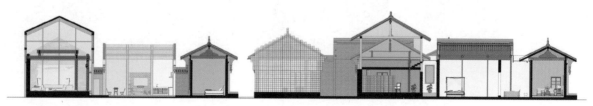

导师点评

Xie Guanyi 谢冠一

  这是一个涉及老城区历史建筑更新活化的设计方案，对于急速发展的中国来说是个时代性的话题。作者区分了待修复的建筑的类型，并统计了街区的人流，设计成果也反应出在现代生活方式与旧建筑保护的关系协调上所做的努力。如果能在调研成果与设计策略之间达成某种逻辑关系，作为支撑最后设计的依据，将会显得更有说服力。版面上内容很多，图文需要合理组织以保证设计重点的表达。目前读图的体验有欠缺。另外，场地调研不应仅局限在建筑测绘，对微观现象的观察和追问有助于揭露隐藏的问题。由于图幅有限，重点内容应适当突出，并删减或整合次要信息，使关注的问题更明确。

This is a design scheme about the renewal and renovation of a historic building in the old city, and it is a historical topic for China, which is booming. The designer makes clear the type of the building to be renovated, and calculates the stream of people in the street. The design results also reflect the efforts made to coordinate the relationship between the modern lifestyle and old building conservation. If a logical relation forms between the research results and design strategy, this logical relation, as the basis of the final design, might make the design more persuasive. There is a lot of design content, so the words and illustrations need to be combined rationally to make the key design points effable. Currently, the illustrations are not so readable. In addition, field research shouldn't be confined to building surveying, while an observation and further investigation of micro-phenomena can help reveal hidden problems. Due to the limited illustration space, key content should be highlighted, while secondary information should be deleted or consolidated, to make the issues of concern clearer

专家点评

Kou Jianchao 寇建超

老城新相——韩城老城历史建筑保护与再利用设计

优点：

  （1）本次设计整体思路充分利用了原有建筑，古城建筑群落布局合理，条理清晰明确。

  （2）景观方面充分尊重中国古代建筑营造法则，体现了人文传承。老城原有建筑的砖墙和木构架被尽量保存，几个重点元素的提取也比较到位，传统与现代结合的设计令古城焕发了新的活力。

  （3）分析图分析清晰，条理明确。

  （4）整体来看排版整齐统一，布局紧凑，分析点突出明确。

缺点：

  （1）色调有点花。背景色和效果图之间没有协调好色调，植物的绿色被衬得过于鲜艳。

  （2）果图的重点不够明确，不能一眼知道效果图要表现或者说设计的重点在哪里。

  （3）文字分析不够到位，分析图可以配合适量文字辅助说明，适量的文字有助于加深观者对图的理解程度。

An Old Town with a New Look—Conservation and Reuse Design of the Historic Buildings in the Ancient City of Hancheng

Strengths:

(1) The overall idea of the design is to make full use of the old buildings. In the buildings in the ancient city are reasonably distributed and well organized.

(2) In terms of landscape, full respect is shown for the construction method of ancient Chinese architecture, reflecting cultural inheritance. The brick walls and timber frames in the old city are well preserved, and several key elements are properly extracted. By combining the traditional with the modern, the design injects new vitality to the ancient city.

(3) The analysis chart is greatly informative, with content in good order.

(4) On the whole, the page layout is neat, uniform and compact, and the analytical points are conspicuous.

Weaknesses:

(1) Heavy color. There is a disharmonious hue between the background color and the design sketch, and the green plants are too bright-colored.

(2) There isn't a clear key point in the design sketch, making it impossible to recognize the design focus at a glance.

(3) The text analysis is insufficient, so in the analysis chart there should be some words as an auxiliary description to deepen readers' understanding of the chart.

恭王府博物馆展览设计·广式家具制作技艺精品展
Exhibition Design for Prince Kung's Museum
——Exhibition on Shanxi-style Furniture Craftsmanship and Masterpieces

「室内设计 6+」2018（第六届）联合毕业设计
"Interior Design 6+"2018(Sixth Year) Joint Graduation Project Event

高　　校：北京建筑大学
College: Beijing University of Civil Engineering and Architecture
学　　生：韩玥、杨玉萍、郑一霖
Students:　Han Yue,Yang Yuping,Zheng Yilin
指导教师：杨琳、陈静勇
Instructors:Yang Lin,Chen Jingyong
参赛成绩：良好
Achievement: Good

学生感悟
Student's Thought

韩玥

"6+"联合毕业设计伴我走完了大学的最后一段路，也是最精彩的一段路。在这段路上，我学习到了很多，在和各高校的交流中，对于设计也有了更深的认识。结交到了许多朋友，留下了一段很美好的回忆。

杨玉萍

"6+"给了我许多了解其他文化的机会，去了许多地方，许多学校，认识许多人，有过许多好的回忆。很开心，学到很多东西。是一段特别特别美好的回忆！值得珍惜！

郑一霖

四年的大学时光即将结束之际，我迎来了"6+"联合毕业设计，而这次答辩可以说是一个圆满的休止符。从严冬到酷暑，终于到了收获成果的时候，在大学的最后半年能与"6+"一同度过，我感到十分荣幸。在这段时间里，我与老师和同学共同合作进步，丰富了更多的技能，彼此都受益匪浅。感谢"6+"给我的这次机会，将会使我在未来的路上更好地前行。

建筑空间设计 | Architectural Space Design

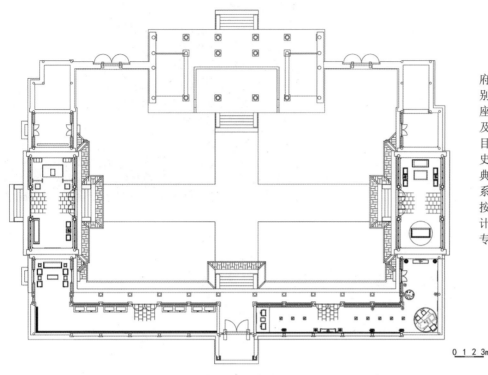

展览区域位于恭王府东二区院落，展馆分别为东西厢房、东西倒座房、东二府门舞台以及院落内外。小组设计目标为梳理广作家具历史脉络与传承人谱系、典型器型与家具形态谱系以及制作工艺谱系，按展馆区域进行展览设计分工，策划广作家具专题展览。

东二区平面图

东二区院落内北立面图　　　　东二区院落内东立面图

导向系统设计 | Guidance System Design

展厅导视牌

院落导视牌

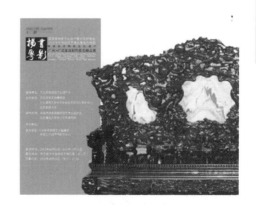

院落外宣传板

由于恭王府院落繁多，行走路线混乱的问题，我们设计了相关导视系统。在导视牌设计中，我们选广作家具中典型器型和纹样做背景，选取王府相关色调，既有王府特色，又能突出展示内容。

西倒座房 | West Reversely-set House

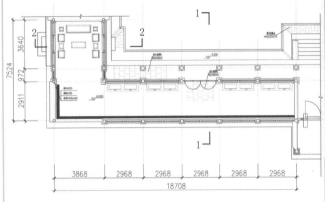

平面图

室内南立面图

西倒座房是南北向开窗，自然采光不充足，我们把正常规格的展柜降低高度，与窗沿平齐，展柜上部采用五面玻璃材质，增加室内空间的亮度。院内根据广州节气种植了节气植物，降低展柜高度后，同时也能从室内欣赏到院内景色。

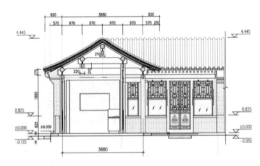

1—1 剖面图　　　　2—2 剖面图

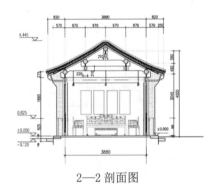

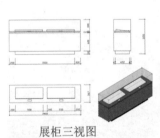

展柜三视图

图签样例

典型器型选择

场景还原部分设置在传承人视频附近,目的是还原清代广州室内的家具陈设,使观展者产生代入感。场景陈设借鉴清代广式家具室内摆法以及杨虾家具店内陈设,陈设物品采用广州特色非物质文化遗产——广绣和广彩。

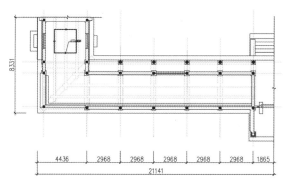

天花图

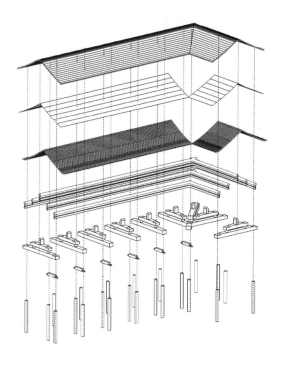

卷棚硬山式屋顶抬梁式构架

国粹赓续卷——历史建筑保护与再利用设计

答辩展示

195

西厢房 | West Wing-room

室外效果图

紫檀高束腰五屏风雕中西纹饰大罗汉床效果图

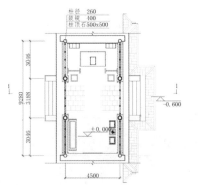
西厢房平面图

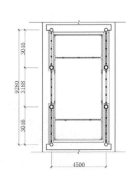
西厢房天花图

西厢房室内东立面图

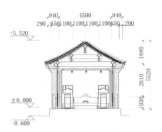
西厢房梁架 1-1 剖面图

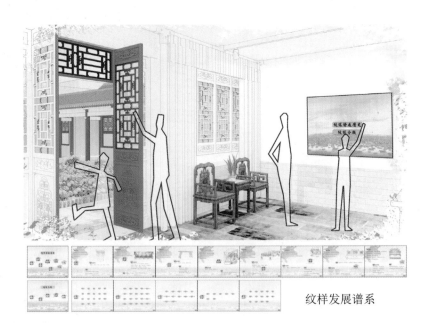
纹样发展谱系

东厢房 | East Wing-room

九龙沙发效果图

梁架结构分析图

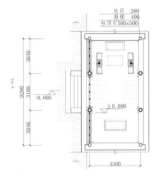

东厢房平面图

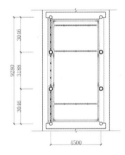

东厢房天花图

东厢房室内西立面图

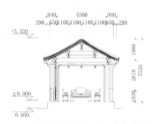

东厢房梁架 2-2 剖面图

此组沙发名为"双凤朝阳",产于名匠居。"名匠居红木家具"成立于20世纪90年代末,至今已有十多年红木家具的生产、销售经验,由国家级非物质文化遗产广式家具制作技艺传承人——杨虾先生创办。凭借其对生产工艺的独到见解及在红木家具制作领域的丰富经验,生产出来的每一件成品都匠心独运,广受消费者喜爱。"双凤朝阳"沙发便是杨虾先生的代表作之一。

雍亲王题书堂深居图屏　桐荫品茶　立持如意　倚门观竹　烘炉观雪

东倒座房 | East Reversely-set House

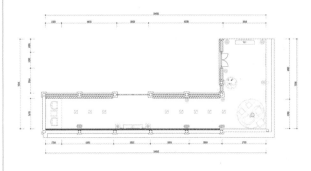

东倒座房平面图

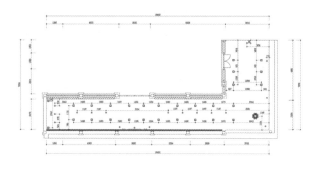

东倒座房天花图

灯光节点分析 | Light Node Design

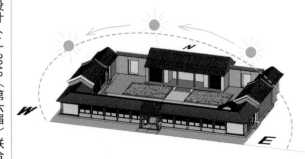

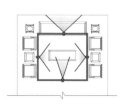

西倒座房灯光分析图

由于东、西倒座房位置原因，两侧开窗朝向南北，自然采光不充足，展厅室内氛围偏暗。于是利用轨道式射灯把灯光打在场景器型和陈设物品上，使观展者聚焦在实物上，更易融入展览氛围。

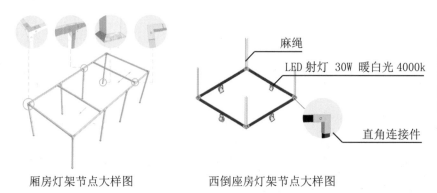

厢房灯架节点大样图　　西倒座房灯架节点大样图

基于古建筑保护原则，我们采用麻绳悬挂灯具。挂镜线受力面积小，对古建筑破坏性相对较大，而麻绳受力面积大，压强小，对古建筑梁架表面破坏性小。

院落设计 | Courtyard Design

在院内引入广东二十四节气特色植物，做到一年之内不同时间景色皆不同。东二府门舞台可作为开幕式场地，平时可以作为粤剧戏台、手工艺演绎场地。

秋分：农历二十四节气中的第十六个节气，时间一般为每年的9月22或23日。

寒露：第十七个节气，时间为每年的10月8日或9日。

立冬：是二十四节气之一，也是汉族传统节日之一，时间为每年11月7-8日之间。

霜降：霜降二十四节气之一，时间为每年的公历10月23日。

导师点评 Wang Yihan 王一涵

"扬书粤影"命题别致，粤影点明设计内容为广式家具制作技艺的展示设计。设计者以历史建筑恭王府为展示设计载体，在对老建筑作为展示设计功能定位分析的基础之上，对场地参观流线、空间功能定位、展示区域划分等进行了重新梳理。将老建筑四合院多处房门封闭，形成内向型相对统一、完整、闭合的参观路线。根据广式家具历史脉络与传承人谱系、典型器型与家具形态谱系及制作工艺谱系划分展览区域，兼顾了历史建筑的保护与展示功能的更新再利用。同时考虑到老建筑的采光问题，在不破坏老建筑的同时，解决了展品照明。整个方案概念表达逻辑清晰，形式与功能相对统一，版面布置有序，成果呈现系统完整。如能在展示设计的多样性与丰富性做一些深入的思考，则整体方案将更富有趣味与创意性。

"Yangshuyueying" is an unconventional topic. Yueying indicates that the design content is the manufacture technology of Guangdong-style house furnishings. The designer chooses historic building Prince Kung's Palace for display. Based on the display design of the old building, the designer re-handles the visiting route, spatial functional localization and display zoning. Several doors in this old courtyard house are sealed off, forming an inward relative uniform, complete and closed visiting route. The interior display space is zoned according to the historical context of Guangdong-style furniture, the pedigree of the inheritors, as well as classic types of ware, furniture modality and fabrication technology, making it possible to protect the historic building and refuse it for displays. Daylighting being taken into account, the problem of illumination is solved on the premise of not destroying the old building. The design concept is expressed in a clear logic, the form and function are relatively in unity with each other, the drawing looks tidy, and the results are presented systematically and completely. The design scheme will look more interesting and creative if the diversity and richness of the display design is further enhanced.

专家点评 Sun Dongning 孙冬宁

本组同学以"扬书粤影"为题，并以国家非物质文化遗产广作硬木家具制作技艺代表性传承人杨虾祖辈的传承谱系为线索，学术梳理广作家具技艺传承内容，并结合动静态展览展示。展览设计包括广作家具典型器形、家具结构、榫卯结构、制作工具等实物展示，历史文脉、传承人谱系、工艺流程、工序及流变。展示设计包括现场制作技艺展示。展演设计包括东二府门舞台设计等内容，在展览期间进行粤剧展演。展板设计内容丰富，视频展示设计包括传承人口述史采访、技艺流程、地域文脉等。本组设计还采用了恭王府历史时期家具陈设场景复原的方式，体现了恭王府活态的文化空间，不仅展示了王府生活的原态，也是在传递、传承优秀的传统文化。

With "Yangshuyueying" as the title, along the inherited pedigree of Yang Xia, a representative inheritor of national intangible cultural heritage Guangdong-style hardwood furniture craftsmanship, the students academically review the inherited content of Guangdong-style furniture craftsmanship and display it both dynamically and statically. The exhibition design involves classic ware of Guangdong-style furniture, furniture structure, tenon-and-mortise structure and tools, as well as historical context, inheritor's pedigree, technological process, production process and flow. The display design involves fabrication on site. The show design involves design of a stage at the east palace gate, and Cantonese Opera will be performed during the exhibition. The display panels have rich content, and video display design involves an interview with the inheritor, concerning technological process and local context showing. Also, the team restores the mode of furnishing in Prince Kung's Palace in history to embody the active cultural spaces in Prince Kung's Palace, including not only the real life in the palace, but also the excellent traditional culture.

有机更新，新旧共生——浙江省薛下庄村『二十四间』老建筑保护与再利用设计研究

Organic Renewal and Co-existence of the Old and the New—Conservation and Reuse Design of "Twenty-four" Old Buildings in Xuexia Village, Zhejiang Province

高　　校：浙江工业大学
College: Zhejiang University of Technology

学　　生：刘叶、蒋一德
Students: Liu Ye, Jiang Yide

指导教师：吕勤智　王一涵
Instructors: Lv Qinzhi Wang Yihan

课题评价：优秀
Achievement: Excellent

学生感悟
Student's Thought

在这次的比赛中，我非常感谢我的老师对我的指导。让我拓宽思路去做自己所感兴趣的设计，如何从设计师的角度去发现问题，解决问题，做一套完整的系统的设计也是本次比赛中巨大的收获。

刘叶

这是一次充满挑战的联合设计，通过一整轮的讨论和方案的推敲，我对老建筑的保护与再利用有了更加深刻的理解和认识，在老师们的辛勤帮助下，收获颇丰。同时，能够与其他院校的老师与同学共同交流，并且得到了评委们的意见与建议，使这次活动变得意义非凡。

蒋一德

二十四间建筑概况 | Overview of Twenty-four Houses

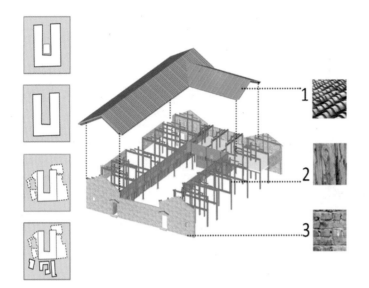

原建筑主要以青砖、木材和瓦为主要材料。

二十四间建筑流线 | Architectural Flow Line of Twenty-four Houses

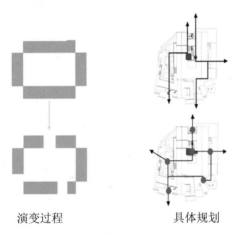

演变过程　具体规划

从原本的单一的从正门到后院的流线改变为多个院落相互联系，形成一个完整的院落体系的流线。

平面图 | Floor Plan

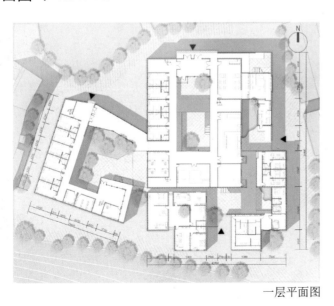

一层平面图

一层平面主要为客房和茶室部分，同时兼顾有厨房等用房，通过各个庭院相互联系在一起。

设计概念分析 | Design Concept Analysis

改造策略 | Renovation Strategies

通过形成新的院落体系，打破传统的院落格局，来实现老建筑的功能更新和再生。

建筑空间生成 | Architectural Space Generation

 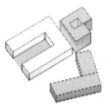  

拆除部分　　　　保留部分　　　　新建部分　　　　连廊部分

通过新建的现代建筑，使用新的形式和新的材料来围合起新的院落体系。

建筑功能生成 | Architectural Function Generation

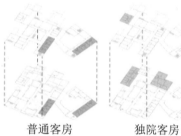 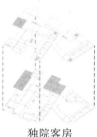 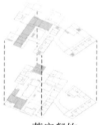 

普通客房　　　　独院客房　　　　茶室餐饮　　　　后勤办公

将客房部分分为普通客房和独院客房，同时在一层安排茶室，二层安排餐饮。

村落鸟瞰 | Bird's Eye View of the Village

室内轴测图 | Interior Axonometric Drawing

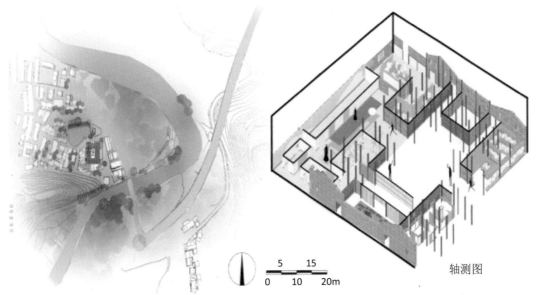

轴测图

室内设计分析 | Interior Design Analysis

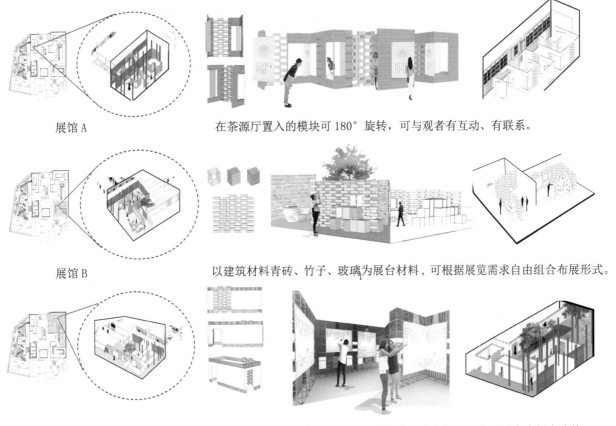

展馆 A　　在茶源厅置入的模块可 180°旋转，可与观者有互动、有联系。

展馆 B　　以建筑材料青砖、竹子、玻璃为展台材料，可根据展览需求自由组合布展形式。

展馆 C　　将模块置入展示空间，每个展示空间有两个门洞，观者可以自由组合流线。

室内设计分析 | Interior Design Analysis

茶室 A

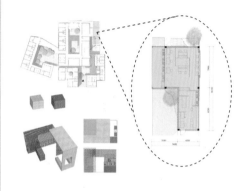

茶室 A 平面图　　　　　茶室 A 效果图

室内的茶座使用混凝土和木材两种材料，从传统到现代，用相互穿插的表达方式，来表达室内饮茶文化，同时与建筑的表达手法相互呼应。

茶室 B

茶室 B 平面图　　　　　茶室 B 效果图　　　　　体块推导过程

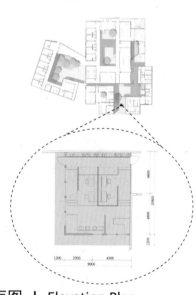

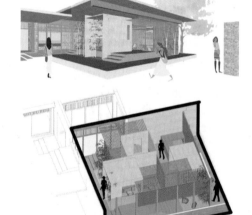

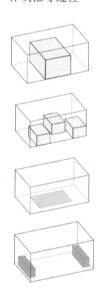

立面图 | Elevation Plan

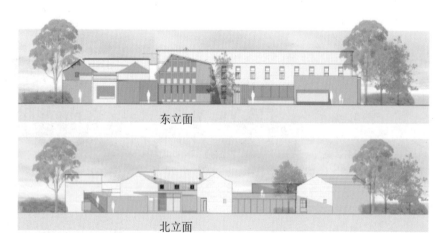

东立面

北立面

造型上通过新与旧建筑的结合，使整个建筑不仅仅是新建筑也不是单纯的老建筑，而是新老更迭的一个生长过程，从而实现建筑的有机更新与新旧共生。

景观分析 | Landscape Analysis

西院景观分析 | Landscape Analysis of West Courtyard

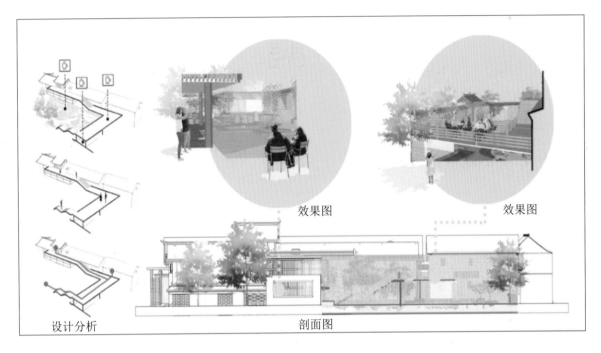

效果图　　　效果图

设计分析　　剖面图

东院景观分析 | Landscape Analysis of East Courtyard

建筑东院的小庭院连接了茶源厅和茶萃厅。结合两个展厅的空间设计，会让观展者有视线上的穿透和流线上的穿插关系。

景观轴测图　　效果图

茶室景观 | Landscape in Teahouse

LOGO 设计 | LOGO Design

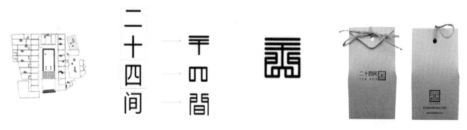

推导过程

周边设计 | Peripherals Design

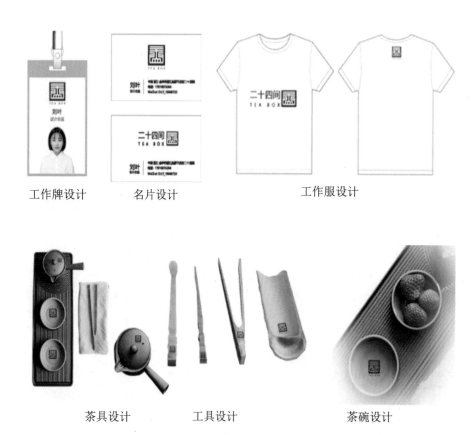

工作牌设计　　名片设计　　　　　工作服设计

茶具设计　　工具设计　　　　茶碗设计

APP 设计 | APP Design

导师点评
刘晓军
Liu Xiaojun

作品薛下庄村"二十四间"老建筑有机更新，其设计者对基地现状调研详实，通过场地空间的分析和梳理提出以保护和再利用为原则的合理规划方向。通过对较完好的老建筑保护修缮，对部分残损建筑拆除重建，使"二十四间"老建筑群更加整体、和谐，充分展现了其地域特征和本土建筑风貌。

改造后的"茶研社"集合了品茶和茶文化展示，室内布局合理聚散富于变化；交通清晰流线丰富，考虑了不同人群的行走需求；功能上充分满足了游客和当地居民的使用；建筑立面的处理上很好地展示了老建筑的结构与肌理；在三间茶厅选用竹子、青砖等乡土材料进行的模块式设计，营造出了一个新颖的茶文化观展体验空间。

作品以乡村振兴为目的对策中把建筑改造和经济文化很好地结合在一起，以旅游带动乡村发展，同时考虑到未来的运营植入了"创客 & 文创产品"的模式，并设计了相关产品，同时采用APP的推广手段，这些策略都具有较高的应用性。

The design involves the organic renewal of "twenty-four" old buildings in Xuexiazhuang Village. After detailed and reliable research on the actuality of the site, the designer makes an analysis of the site space and puts forward a rational plan for the protection and reuse of the buildings. After conservation and renovation of some well-preserved old buildings and reconstruction of some damaged buildings, the "twenty-four" old buildings are combined together and stand in harmony with one another, fully displaying their regional characteristics and the local architectural scene.

After transformation, the "tea research club" is used for tea tasting and tea culture display, and the indoor layout is reasonable and changeful; the traffic is clear and greatly streamlined, with different people's needs for walking indoors taken into account; functionally, tourists' and the local residents' requirements for utilization are fully met; after treatment, the building facade well displays the structure and texture of the old buildings; in the three tea stalls, bamboo, black bricks and other local materials are used for modular design, creating a novel space where tea culture can be observed and displayed.

To achieve the goal of rural revitalization, the designer well combines building renovation with the economy and culture, so that tourism could drive the rural development. Moreover, considering future operation, the mode of "maker & cultural creative products" is adopted, and related products are designed. Then, the products are promoted by APP, and these strategies are highly applicable.

专家点评
王传顺
Wng Chuanshun

"二十四间"老建筑保护与再利用设计研究，作品以有机更新、新旧共生为重点，通过深入调查研究、分析，结合原建筑院落系统，将新建的现代建筑，使用新的形式和新的材料来围合起新的院落体系，使用空中连廊和屋顶平台的形式，增加二层的空间和一层的院落体验，充分体现出对古建筑的利用保护及新建筑表达融合。室内使用的混凝土和木材等，从传统到现代，相互穿插，为村民和游客两种不同的人群提供了一个交流的平台。

本设计从规划、建筑、景观、室内一体化考虑设计，作品主题明确，反映出具有较好的专业基本功和实践应用能力，打造出一个完美的村落。

The project is about conservation and reuse design of "twenty-four" old houses. The group, focused on organic renewal and co-existence of the old and the new, builds a new courtyard system with new modern buildings and materials in a new form after in-depth research and analyses based on the original courtyard system. A space corridor and roof terrace is built, enlarging the second floor space and the courtyard in the first floor, giving full expression to the utilization and conservation of ancient buildings and the fusion of ancient buildings with new buildings. Concrete and wood are used indoors, with the traditional and the modern mixed together, providing the villagers and tourists, two different groups, a platform to communicate with each other.

The design is made from the overall perspective of planning, construction, landscaping and interior decoration, and the work is given a clear-cut theme, revealing the designer's good professional foundation and practical ability, with which a perfect village is built.

热点命题,纷显特色

联合指导,服务需求

教育研究

室内设计6+ 2018（第六届）
联合毕业设计
Interior Design 6+ 2018 (Sixth Year)
University-Enterprise Cooperative Graduation Project Event

中国建筑学会室内设计分会《"室内设计6+"联合毕业设计特色教育创新项目》报告
(2013届～2017届)

为服务城乡建设领域室内设计专门人才培养需求，加强室内设计师培养的针对性，促进相关高等学校在专业教育教学方面的交流，引导面向建筑行（企）业需求开展综合性实践教学工作，由中国建筑学会室内设计分会（以下简称室内分会）倡导、主管，国内外设置室内设计相关专业（方向）的高校与行业代表性建筑与室内设计企业开展联合毕业设计项目。

历经2013～2017连续5届联合毕业设计项目的深入交流，取得了丰富成果，形成一定影响力，积累了室内分会设计教育平台建设成功经验。现对前5届项目开展情况做出总结。

一、特色教育创新项目设立的背景、目的和意义

2010年教育部启动了"卓越工程师教育培养计划"，于2011年-2013年分三批公布了进入"卓越计划"的本科专业和研究生层次学科。2011年国务院学位委员会、教育部公布《学位授予和人才培养学科目录（2011年）》，增设了"艺术学（13）"学科门类，将"设计学（1305）"设置为"艺术学"学科门类中的一级学科。"环境设计"建议作为"设计学"一级学科下的二级学科，"室内设计"建议作为新调整的"建筑学（0813）"一级学科下的二级学科。2012年教育部公布《普通高等学校本科专业目录（2012年）》，在"艺术学"学科门类下设"设计学类（1305）"专业，"环境设计（130503）"等成为其下核心专业。"艺术学"门类的独立设置，设计学一级学科以及环境设计、室内设计等学科专业的设置与调整，形成了我国环境设计教育和室内设计专门人才培养学科专业的新格局。

2015年10月，国务院发布《统筹推进世界一流大学和一流学科建设总体方案》。2017年1月，教育部 财政部 国家发展改革委印发《统筹推进世界一流大学和一流学科建设实施办法（暂行）》。党的十九大报告指出："建设教育强国是中华民族伟大复兴的基础工程，必须把教育事业放在优先位置，加快教育现代化，办好人民满意的教育。""一流大学和一流学科建设"是建设高等教育强国、实现十九大提出的"实现社会主义现代化和中华民族伟大复兴"总任务的必然选择和重要举措。

因此，组织开展室内设计领域联合毕业设计，对加强相关学科专业特色建设，深化综合性实践各教学环节交流，促进室内设计教育教学协同创新，培养服务行（企）业需求的室内设计专门人才，具有十分重要的意义。

二、2013届～2017届联合毕业设计项目开展情况

Report on the Innovation Program of Characteristic Education on "Interior Design 6+" Joint Graduation Project
(2013~2017)

To meet the demand for professional interior design talent training in the urban and rural construction field, strengthen the pertinence of interior designer cultivation, promote related universities to exchange ideas with one another on professional education and teaching, and assist the universities in carrying out comprehensive practice teaching to meet the needs of the construction industry (enterprises), the domestic and foreign universities that offer interior design-related specialties and representative architectural and interior design enterprises cooperate in doing the graduation project on the initiative of the Institute of Interior Design-ASC (IID-ASC).

After an in-depth exchange of views on the joint graduation project event in the past 5 years from 2013 to 2017, great achievements have been made, accumulating successful experience in the construction of an educational platform for interior design. The following is a summary of the previous 5 years' achievements.

I. The Background, Objective and Significance of the Characteristic Educational Innovation Program.

In 2010, the Ministry of Education initiated the "Excellent Engineer Training Program"; in 2011-2013, it made public the undergraduate specialties and postgraduate programames listed in the "Excellent Program" in three times. In 2011, the Academic Degree Commission of the State Council and the Ministry of Education released the List of the Degree and Talent Training Subjects (2011), additionally offering the "art (13)" specialty, setting the "design (1305)" as a first-level subject in the "art" specialty. The "environment design" is proposed as a second-level subject subordinate to the "design", and the "Interior design" is proposed as a second-level subject subordinate to the "architecture (0813)", which had just been upgraded to be a second-level subject. In 2012, the Ministry of Education released the Catalogue of the Undergraduate Programs offered at Regular Institutions of Higher Education (2012), putting the "design (1305)" and "environment design" under the category of the "art" as core specialties. The setup and adjustment of the "art", as well as the design, a first-level subject, and the interior design formed a new pattern of professional interior design talent training in the area of environment design education.

In October 2015, the State Council issued An Overall Plan on Comprehensively promoting the Construction of World-class Universities and Subjects. In January 2017, the Ministry of Education, the Ministry of Finance and the National Development and Reform Commission printed and issued Measures for Comprehensively promoting the Construction of World-class Universities and Subjects (Interim). The report of the 19th CPC National Congress indicates: "It is a foundation project related to the rejuvenation of China to build China into a great power of education. This necessitates making the educational cause a priority, quickening educational modernization and meeting the people's requirements for education." The "construction of world-class universities and subjects" is an inevitable choice and important measure to build China into a great power of higher education and fulfill the overall task of "realizing socialist modernization and the rejuvenation of China" set at the 19th CPC National Congress.

Therefore, joint graduation project of interior design is of

1. 项目指导单位、主办单位

"室内设计6+"联合毕业设计受全国高等学校建筑学学科专业指导委员会、教育部高等学校设计学类专业教学指导委员会等指导，由室内分会主办。

2. 项目基本设置与定位

《"室内设计6+"联合毕业设计特色教育创新项目》连续5届的活动，有6+1所区域高校、6+1个校企联合、6+1个项目环节、6+1个编制栏目、6+……，使"6+"成为室内分会这个室内设计教育示范项目成为品牌平台，彰显出不断丰富的新内涵和不断提升的影响力。

项目坚持联合企业开展联合毕业设计教学，体现服务于人才培养行业需求的原则。经参加高校自主联系并向学会备案，探索形成了在总命题《框架任务书》指导下的6+1所高校、6+1家企业、6+1处特色建筑的"6+"模式联合毕业设计项目的新方式，突显出项目"热点命题，纷显特色，联合指导，服务需求"的新定位。

3. 项目实施（核心）高校

联合毕业设计一般由学科专业条件相近，设置室内设计方向的相关专业的6所高校间通过协商、组织成为项目参加高校。应突出参加高校组合的地理区域、办学类型、专业特色、就业面向等的代表性、涵盖性，在学科专业间形成一定的交叉性和协同设计工作环境和交流氛围。

在建设创新型国家和人才强国战略的指引下，立足我国学科专业的新环境，面对建筑室内设计人才培养的新目标，服务行业发展的新需求，2013年，由室内分会主办，由同济大学、华南理工大学、哈尔滨工业大学、西安建筑科技大学、北京建筑大学、南京艺术学院等6所发起高校（2015年加浙江工业大学后成为6+1所），地处不同区域、设置建筑学或设计学等学科室内设计方向的高校与知名设计企业，共同创建了室内设计教育示范项目："室内设计6+"联合毕业设计，联合企业协同探索"卓越计划"目标下的室内设计师教育培养之路。

每所高校参加联合毕业设计到场汇报的学生一般以6人为宜，分为2个设计方案组；要求配备1～2名指导教师，其中至少有1名指导教师具有高级职称；高校导师熟悉环境设计、室内设计等工程实践业务，与相关领域企业联系较广泛。室内分会负责聘任参加高校导师，参与联合毕业设计工作。

4. 项目命题企业

参加高校向室内分会推荐所在省（市、地区）的行业代表性建筑与室内设计企业作为毕业设计命题企业，企业命题人应具有高级职称；室内分会负责聘任企业命题人作为联合毕业设计企业导师。企业导师与相应高校导师协同编制联合毕业设计总命题下的《"（分课题）×"毕业设计教学任务书》，

great significance to strengthening the characteristic construction of relevant subjects and specialties, deepening exchanges on comprehensive practical teaching, promoting collaborative innovation of interior design education and teaching, and training professional interior design talents required for service industries (enterprises).

II. Status of the Joint Graduation Project Event in 2013~2017

1. Project Guider and Organizer

The "Interior Design 6+" Joint Graduation Project Event is organized by the IID-ASC, and guided by the Professional Guiding Committee for Architecture at Chinese Universities, the Ministry of Education and other various professional design teaching guidance committees.

2. Project Settings and Positioning

The Characteristic Educational Innovation Program on "Interior Design 6+" Joint Graduation Project, which has been organized for 5 consecutive years, has 6+1 participants, which are regional universities, 6+1 university-enterprise joint groups, 6+1 project segments, 6+1 columns, 6+…, making the "6+" interior design education demonstration project a brand platform, demonstrating a constantly enriched new connotation and continuously improved influence.

Perseverance in implementing joint graduation project teaching reflects the dedication to serving talent training to meet industry needs. Every university involved needs to apply to the IID-ASC, participate in the "6+" joint graduation project event under the guidance of the Framework Assignment. This highlights the new positioning of the project that it "concerns a hot topic, demonstrates characteristic joint guidance and meets demands".

3. (Core) Project Implementers (Universities)

The joint graduation project event is generally attended by 6 universities that offer similar subjects and specialties, including the interior design, through consultation. These universities' geographic region, teaching type, professional characteristics and employment orientation should be highlighted, to hold different disciplines together, and create an environment and atmosphere of communication for collaborative design.

Guided by the strategy of making China an innovative nation and vitalizing China with human power, based in the new subject and specialty environment of China, faced with the new objective of interior design personnel training, and betaken to meeting the new demand of the industry development, Tongji University, South China University of Technology, Harbin Institute of Technology, Xi'an University of Architecture and Technology, Beijing University of Civil Engineering and Architecture and Nanjing University of the Arts, which are located in different regions and offer the same architectural interior design, initiated an interior design education demonstration project in collaboration with well-known design enterprises under the auspices of the CIID in 2013. Then in 2015, Zhejiang University of Technology joined the project, making it a "6+1" project, named "Interior Design 6+" Joint Graduation Project Event. The universities join hands with the enterprises to explore a way to educate and design interior designers under the goal of the "Excellent Program".

For each university involved in the joint graduation project event, it sends 6 students, divided into 2 teams, to give a report on the spot; 1~2 supervisors, including at least one with a senior professional title, take part in the event with the students; university supervisors are familiar with engineering practice, including environment design and interior design, and stay in

参与联合毕业设计相关环节工作。

5. 项目活动命题

联合毕业设计选题始终是以行业、社会密切关注的热点问题为导向。2013届～2017届联合毕业设计项目的各届命题和涉及热点问题分别是：

（1）2013（首届）的"赛后商机——国家体育场赛后改造室内设计"课题，涉及了大型体育赛事之后大型体育场馆经营利用问题。

（2）2014（第二届）的"城轨新境——上海地铁改造环境设计"课题，探讨了城市更新与地铁站点的环境设计问题。

（3）2015（第三届）的"兵工遗产——南京晨光1865创意产业园环境设计"，研究了历史文化名城近代工业遗产保护与利用问题。

（4）2016（第四届）的"鹤发医养——北京曜阳国际老年公寓环境改造设计"，挖掘了既有医养建筑空间环境适应老龄社会宜居新需求问题。

（5）2017（第五届）的"国匠承启——传统民居保护性利用设计"，比较了多地域代表性民居保护和利用的问题。

6. 项目支持企业

参加高校向室内分会推荐行业代表性建筑与室内设计企业作为毕业设计支持企业；由室内分会择优与支持企业签订项目支持与回馈协议，负责聘任支持企业观察员，参与联合毕业设计相关环节工作。

7. 成果出版企业

室内分会和《主题卷》总编高校遴选行业知名出版企业，作为联合毕业设计《主题卷》出版企业，并参与联合毕业设计相关环节工作。

8. 项目成果出版物

将每届项目成果梳理形成了室内分会推荐设计教学参考书，也初步形成了"室内设计6+"联合毕业设计教学丛书的编制文化和书籍设计风格。每卷的6+1个"编制栏目"涵盖了相应"项目环节"的核心内容。其中，"调研踏勘""中期检查""答辩展示""教育研究""专家讲坛""历史定格"6个栏目记录了当届项目的主体和过程内容，还附有专家点评、学生感言、获奖证书、师生照片等；经过5届不断完善、优化形成的项目《章程》、本届《项目纲要》《框架任务书》《答辩、评审、表彰工作细则》等，汇编在第1栏目"项目规章"中，是指导项目开展的长效机制，是室内分会室内设计教育工作有章可循的体现。

三、联合毕业设计项目流程和环节

1. 联合毕业设计每年举办1届，与参加高校毕业设计教学工作实际相结合。

2. 室内分会负责联合毕业设计总体策划、宣传、组织研讨、编制、公布每届联合毕业设计《（主题）×——（总命题）× 框架任务书》《项目纲要》，制订、公布《"室内设计6+"联合毕业设计答辩、

touch with many enterprises concerned. The IID-ASC is obliged to engage supervisors from the universities involved and have them participate in the joint graduation project event.

4. Topic Assigner

Each university involved shall recommend to the IID-ASC a representative provincial (municipal or regional) interior design enterprise as the topic assigner, and the enterprise shall have a senior professional title; the IID-ASC shall hire the topic assigner as a corporate supervisor of the joint graduation project event. The corporate supervisor shall assist the university supervisor(s) concerned in compiling a "(sub-topic)×" graduation project teaching assignment under the General Assignment, and guiding the joint graduation project event.

5. Project Topic Assignment

The joint graduation project event is always directed at hot issues closely watched by the insiders and the society. The topics and hot issues involved of 2013~2017 are as follows:

(1) 2013 (the first year): "Post-game Business Opportunities—Post-game Interior Renovation Design of National Stadiums", involving the management and utilization of large-scale stadiums after the end of major sports events.

(2) 2014 (the second year): "A New Look of Urban Rails—Environmental Renovation Design of the Shanghai Metro", involving urban renewal and environmental design of metro stations.

(3) 2015 (the third year): "Heritage of Military Engineering—Environmental Design of Nanjing Chenguang 1865 Creative Industry Park", involving the conservation and utilization of the modern industrial heritage in the historic city.

(4) 2016 (the fourth year): "Pension Healthcare—Environmental Renovation Design of Beijing Yaoyang International Apartments for the Elderly", involving the use of existing healthcare architectural space environments to meet the new needs of old age support.

(5) 2017 (the fifth year): "Inheritance of National Craftsmanship—Protective Utilization Design of Traditional Dwellings", involving a comparison of protection and utilization among the representative dwellings of many regions.

6. Corporate Supporter

Each university involved shall recommend to the IID-ASC a representative architectural and interior design enterprise as a supporter for the joint graduation project event; the IID-ASC shall sign a event support and feedback agreement with the supporter and hire it as an inspector to participate in the joint graduation project event.

7. Achievements Publisher

The IID-ASC and the chief compiler of the Topic Volume shall select a well-known publishing enterprise as the Topic Volume and a participant in the joint graduation project event.

8. Achievements Publication

The project achievements of each year has been compiled into a design teaching reference book recommended by the IID-ASC, initially forming the culture of compilation of teaching books for "Interior Design 6+" and the design style of books. The 6+1 "columns" in each volume cover the core content of the whole "project process". Of the columns, "on-the-spot survey", "in-process inspection", "oral defense display", "education research", "expert forum" and "history freeze-framing" record the subject and process content of each year's project, and attach expert comments, students' words, honor certificates and teacher-

评价工作细则》等，协调参加高校、命题企业、相关机构等，聘请课题方向领域专家开展专题学术讲座，组织对毕业设计课题成果质量、毕业设计优秀指导教师、毕业设计优秀组织单位、毕业设计特殊贡献等的评价，以及室内设计教育国际交流等。

3. 联合毕业设计主要教学环节包括：命题研讨、开题报告、中期检查、答辩评审及课题评价、《主题卷》编辑出版、专题展览等6主要环节，对外交流作为联合毕业设计项目的1个扩展环节。相关工作分别由室内分会、参加高校、命题企业、支持企业、出版企业等分工协同落实。

4. 命题研讨

室内分会组织召开联合毕业设计命题研讨会。每届联合毕业设计的总命题着眼室内设计等相关领域学术前沿和行业发展热点问题，参加高校协同命题企业细化总命题方向下分课题。联合毕业设计分课题要求具备相关设计资料收集、现场踏勘、建设管理方支持等条件。

命题研讨会一般安排在高校秋季学期中（每年11月左右），结合当年室内分会年会安排专题研讨。

5. 开题报告

室内分会组织开展联合毕业设计开题活动，颁发联合毕业设计高校导师和企业导师聘书，承办高校协同安排开题活动启动、专家学术讲座、开题汇报与专家点评、调研参观等。每所参加高校合组报告开题情况，其中分命题介绍与开题报告陈述不超过20分钟，专家点评不超过10分钟。

开题活动一般安排在高校春季学期开学初（3月上旬）进行。

6. 中期检查

室内分会组织开展联合毕业设计中期检查活动，承办高校协同安排专家学术讲座、中期成果汇报与专家点评、调研参观等。每所参加高校优选不超过2个过程方案组进行中期检查，其中每组陈述不超过20分钟，专家点评不超过10分钟。

中期检查一般安排在春季学期期中（4月下旬）进行。

7. 答辩评审与课题评价

室内分会组织开展联合毕业设计答辩评审及项目评价，承办高校协同安排答辩评审、课题评价等工作。每所参加高校优选不超过2个答辩方案组进行陈述与答辩、成果展出，其中每组陈述不超过20分钟，专家点评与学生回答不超过10分钟。

在答辩、成果展示、评审的基础上，室内分会组织开展对《"室内设计6+"联合毕业设计特色教育创新项目》的年度课题评价，重点评价毕业设计课题成果质量、毕业设计优秀指导教师、毕业设计优秀组织单位、毕业设计特殊贡献等。坚持"博采众议、考教分离 质量第一、宁缺毋滥"的评审原则，毕业设计课题成果质量评价结果分为优秀、

student photos; after 5 years of constant improvement, the Project Charter, as well as this year's Project Outline, Framework Assignment and Detailed Regulations on Oral Defense, Review and Commendation has taken shape. The compilation, which is in the 1st column Project Charter, is a long-term mechanism under which the project is carried out, and the embodiment of that the IID-ASC has rules to follow while developing interior design education.

III. Joint Graduation Project Event Procedures

1. The join graduation project event, held once a year, is in line with the universities' graduation project teaching.

2. The IID-ASC shall make overall planning for the joint graduation project event, publicize it, organize discussions, compile files, release the (Subject)×—(General Assignment)×Framework Assignment and Event Outline, make and release Guidelines on "Interior Design 6+" Joint Graduation Project Defense and Evaluation", and so on. Also, it shall assist the universities, topic assigners and other agencies in hiring professional experts to hold an academic forum, and organize a evaluation on the quality of the design results, excellent graduation project supervisors, excellent graduation project participants and special contributions to the graduation project event, as well as international communication on interior design education.

3. The process of joint graduation project teaching consists of 6 parts: topic assignment discussion, in-process inspection, defense review and topic evaluation, compilation and publication of the Topic Volume, and special exhibition, while external communication is an extension of the joint graduation project event. The above work shall be done by the IID-ASC, universities, topic assigners, supporters and publishing enterprises separately or synergistically.

4. Topic Assignment Discussion.The IID-ASC shall organize a discussion on topic assignment for the joint graduation project event. The general assignment of each year's joint graduation project event shall be focused on the academic frontier and industrial hot spots of interior design, and the universities involved shall make a subtopic under the general assignment. The subtopic of the joint graduation project event requires design data acquisition, reconnaissance trip, and support from construction management.

Generally, a topic assignment discussion is held around the annual topic in the middle autumn term (about November every year).

5. Opening Report.The IID-ASC shall organize an opening report meeting on the joint graduation project event, award a letter of appointment to university supervisors and corporate supervisors assist in holding a academic forum, making opening speeches and comments, and organize research and visits. Every university shall give one opening report, in which subtopic introduction and opening report presentation shall take less than 20min, and expert comments shall take less than 10min.

Generally, an opening report is given in the beginning of the spring term (early March).

6. In-process Inspection.The IID-ASC shall organize an in-process inspection on the joint graduation project event, to assist university experts in holding an academic forum, report in-process achievements, and organize research and visits. Every university involved shall choose up to 2 process scheme teams for in-process inspection, and either team shall give a presentation that takes

良好、合格、不合格四个等级，其中优秀、良好评价结果一般按照1:2比例设置。毕业设计课题成果质量评价仅针对答辩方案设置，评价结果等级可以空缺。

答辩评审及课题评价一般安排在春季学期期末（6月上旬）进行。

8. 专题展览

室内分会在每届联合毕业设计结束当年的室内分会年会暨学术研讨会（每年10～11月份）举办期间安排联合毕业设计作品专题展览；专题展览结束后，相关高校可自愿向室内分会申请联合毕业设计作品巡回展出。

9. 编辑出版

基于每届联合毕业设计成果，由室内分会组织编辑出版《主题卷》，作为室内分会推荐的专业教学参考书。《主题卷》总编工作由室内分会和总编高校、参编高校联合编著，参加高校导师负责本校排版稿的审稿等工作，出版企业作为责任编辑，负责校审、出版、发行等工作。

基于2013届～2017届联合毕业设计成果，每年由室内分会和北京建筑大学联合主编出版了"中国建筑学会室内设计分会推荐设计教学参考书"共5卷：

（1）中国建筑学会室内设计分会编．赛后商机［卷］：国家体育场赛后改造室内设计——CIID"室内设计6+1"2013（首届）校企联合毕业设计［M］，北京：中国水利水电出版社，2013．

（2）中国建筑学会室内设计分会编．城轨新境［卷］：上海地铁改造环境设计——CIID"室内设计6+1"2014（第二届）校企联合毕业设计［M］，北京：中国建筑工业出版社，2014．

（3）中国建筑学会室内设计分会编．兵工遗产［卷］：南京晨光1865创意产业园环境设计——CIID"室内设计6+1"2015（第三届）校企联合毕业设计［M］，北京：中国水利水电出版社，2015．

（4）中国建筑学会室内设计分会编．鹤发医养［卷］：北京曜阳国际老年公寓环境改造设计——CIID"室内设计6+1"2016（第四届）校企联合毕业设计［M］，北京：中国水利水电出版社，2016．

（5）中国建筑学会室内设计分会编．国匠承启［卷］：传统民居保护性利用设计——CIID"室内设计6+1"2017（第五届）校企联合毕业设计［M］，北京：中国水利水电出版社，2017．

not more than 20min, and each expert shall spend not more than 10min to make comments.

Generally, in-process inspection is conducted in the middle spring term (late April).

7. Defense Review and Topic Evaluation. The IID-ASC shall organize a defense review and topic evaluation activity for joint graduation project design in collaboration with the universities involved. Every university involved shall choose up to 2 defense scheme teams for presentation, oral defense and a display of achievements. To be specific, either team shall give a presentation of up to 20min; expert comments and student answers shall take up to 10min.

On the basis of defense, achievement display and review, the IID-ASC shall organize an annual evaluation on the topic of the Characteristic Educational Innovation Program on "Interior Design 6+" Joint Graduation Project, to primarily evaluate the quality of the design results, as well as excellent graduation project supervisors, excellent graduation project participants and special contributions to graduation project design. Review shall follow the principle of "quality first and quality superior to quantity". The results of the evaluation on the quality of the topic achievements are divided into four levels: excellent, good, qualified and unqualified, of which the ratio of excellent to good is 1:2. The quality evaluation is for thesis defense only, and the real evaluation results may not necessarily at all the four levels.

Generally, defense review and topic evaluation are performed in the late spring term (early June).

8. Special Exhibition. The IID-ASC organizes a special exhibition of joint graduation project contents after the joint graduation project event ends and during the period of the annual meeting and academic conference; after the end of the special exhibition, the universities concerned can voluntarily apply to the IID-ASC for an itinerant exhibition of joint graduation project contents.

9. Compilation and Publication. The IID-ASC organizes the compilation and publication of the Topic Volume based on each year's joint project results as a professional teaching reference book. The Topic Volume shall be co-compiled by the IID-ASC, the chief compiler and the other universities involved, and each university's supervisors shall be responsible for examining its own manuscript, while the publishing enterprise shall serve as an editor in charge for proofreading and publication.

Based on the joint graduation project results of 2013~2017, the CIID and Beijing University of Civil Engineering and Architecture compile 5 volumes of "design teaching reference books recommended by the CIID" each year:

(1) Post-game Business Opportunities [Volume]: Post-game Interior Renovation Design of National Stadiums—CIID "Interior Design 6+1" 2013 (First Year) University-Enterprise Joint Graduation Project Event [M], compiled by the CIID. Beijing: China WaterPower Press, 2013.

(2) A New Look of Urban Rails [Volume]: Environmental Renovation Design of the Shanghai Metro—CIID "Interior Design 6+1" 2014 (Second Year) University-Enterprise Joint Graduation Project Event [M], compiled by the CIID. Beijing: China Architecture & Building Press, 2014.

(3) Heritage of Military Engineering [Volume]: Environmental Design of Nanjing Chenguang 1865 Creative Industry Park—CIID "Interior Design 6+1" 2015 (Third Year) University-Enterprise Joint Graduation Project Event [M], compiled by the CIID. Beijing: China WaterPower Press, 2015.

(4) Pension Healthcare [Volume]: Environmental Renovation

10. 对外交流

室内分会和出版企业一般在每届联合毕业设计结束当年室内分会年会期间联合举行《主题卷》发行式；由室内分会联系亚洲室内设计联合会（AIDIA）等，开展室内设计教育成果国际交流，宣传中国室内设计教育，拓展国际交流途径。

四、联合毕业设计项目相关经费

1. 室内分会负责筹措对毕业设计课题成果质量、毕业设计优秀指导教师、毕业设计优秀组织单位、毕业设计特殊贡献等的评价经费，以及室内分会年会专题展览、宣传等环节经费。

2. 参加高校自筹参加联合毕业设计相关师生各环节经费。

3. 承办高校负责联合毕业设计开题报告、中期检查、答辩评审与课题评价等环节的宣传海报、场地、设备等；答辩评审环节承办高校还负责答辩作品展出场地、展板制作等经费；《主题卷》总编高校负责相应出版主要经费等。

4. 命题企业、支持企业、出版企业等负责为向校企联合毕业设计提供一定形式的支持等。

总结和展望

历经 2013～2017 连续五届联合毕业设计的深入交流，原 CIID "室内设计 6+1" 校企联合毕业设计取得了丰富成果，形成一定影响力，积累了室内设计分会设计教育平台建设成功经验。2017 年 10 月，室内设计分会第八届理事会通过《教育工作规划纲要（2018 年 -2025 年）》，将该活动提升为今后持续开展的《"室内设计 6+" 联合毕业设计特色教育创新项目》，更名为 "室内设计 6+" 联合毕业设计。

发挥《"室内设计 6+" 联合毕业设计特色教育创新项目》在 "双一流建设" 中促进作用，可以根据相关代表性特色高校条件及其组合意愿情况，逐步在我国六大区（华北、华东、中南、东北、西南、西北地区）增设（×地区）"室内设计 6+" 联合毕业设计，在各省市增设（×省 / 市）"室内设计 6+" 联合毕业设计。

未来室内设计分会将继续开展 "室内设计 6+" 联合毕业设计教学特色创新（示范）项目，作为分会设计教育交流品牌项目，形成新一轮示范辐射带动。

中国建筑学会室内设计分会
2018 年 4 月 20 日

Design of Beijing Yaoyang International Apartments for the Elderly—CIID "Interior Design 6+1" 2016 (Third Year) University-Enterprise Joint Graduation Project Event [M], compiled by the CIID. Beijing: China WaterPower Press, 2016.

(5) Inheritance of National Craftsmanship [Volume]: Protective Utilization Design of Traditional Dwellings—CIID "Interior Design 6+1" 2017 (Third Year) University-Enterprise Joint Graduation Project Event [M], compiled by the CIID. Beijing: China WaterPower Press, 2017.

10. External Exchanges.Generally, the IID-ASC and publishing enterprise co-distribute the Topic Volume after the end of the joint graduation project event and during the IID-ASC's annual meeting; the IID-ASC shall communicate with the Asia Interior Design Institute Association (AIDIA) on the results of interior design education, propagandize Chinese interior design education and expand the way of international communication.

IV. Costs of Joint Graduation Project

1. The IID-ASC is responsible for raises funds for evaluation on the quality of the design results, as well as excellent graduation project supervisors, excellent graduation project participants and special contributions to graduation project design, as well as for the special exhibition and propaganda.

2. Costs of universities' participation in the joint graduation project event.

3. The universities shall prepare posters, spaces and equipment for the opening report, in-process inspection, oral defense review and topic evaluation; also, the universities shall fund the display of exhibition and the production of display panels; the chief compiler of the Topic Volume shall fund the publication.

4. The topic assigner, supporter and publishing enterprise shall offer some support to the joint graduation project event.

Conclusions and ProspectsAfter an in-depth exchange of views on the joint graduation project event in the past 5 years from 2013 to 2017, the former "Interior Design 6+1" University-enterprise Joint Graduation Project Event achieved fruitful results, accumulating successful experience in the construction of an educational platform for interior design. In October 2017, the 8th council of the IID-ASC passed the Educational Planning Framework (2017-2025), upgraded the event as Characteristic Educational Innovation Program on "Interior Design 6+" Joint Graduation Project, renaming it "Interior Design 6+" Joint Graduation Project Event.

To give full play to the role of the Characteristic Educational Innovation Program on "Interior Design 6+" Joint Graduation Project in "double first-class construction", the (×region) "Interior Design 6+" Joint Graduation Project Event and China (×province/city) "Interior Design 6+" Joint Graduation Project Event can be carried gradually out in the six regions of China (North China, East China, South Central China, Northeast China, Southwest China and Northwest China) and all provinces and cities respectively in accordance with related representative characteristic universities' conditions and will to combine with one another.

In the future, the IID-ASC will continue to implement "Interior Design 6+" Joint Graduation Project Teaching Characteristic Innovation (Demonstration) Project, which will serve as an educational exchange brand project to help form a new demonstrative and radiant effect.

Institute of Interior Design-ASC (IID-ASC)
Submitted to the Architectural Society of China on April 20th, 2018

2018同济大学图书馆保护与更新改造毕设指导心得

左琰　林怡

由中国建筑学会室内分会主办的"室内设计6+"2018（第六届）联合毕业设计的主题为"国粹赓续——历史建筑保护与再利用设计"，同济大学特别选择了位于上海四平路主校区内最具校园记忆和历史风采的建筑之一——同济大学图书馆的保护与更新改造作为今年的毕设课题，一方面是让学生对校园文化和校园历史建筑的保护价值有一个基本的认知和态度；另一方面也是希望学生深入调研后归纳和提炼主要现状问题，提高学生整合设计能力和解决实际问题的能力，特别在光环境上探索优化改进措施，实现由"面"到"点"的逐级深入，提高方案落地性。

1. 图书馆的历史变迁和价值解读

位于上海四平路主校区内的同济大学图书馆始建于20世纪60年代，与同济校门、南北楼和毛主席像等建于同时期，形成了校园历史中轴线上一个重要建筑景观。图书馆落成使用至今已有50多年，已经历过三次改扩建。始建的图书馆为两层高的砖混结构，外立面为红砖墙，空间呈对称"日"字型院落布局，这种空间模式与当时的图书馆使用方式相关，紧凑的三层书库为非开放区域，借还书需通过管理人员操作。

第一次扩改建为20世纪80年代，由于用地限制，在原有中央内院中建立两个悬挑塔楼以解决阅览空间过少的问题。塔楼结构采用时兴的钢筋混凝土核心筒悬挑结构，出挑接近10m，结构创新在当时具有引领意义，共11层高的塔楼立面采用与红砖墙相近的马赛克贴面，扩建完的建筑无论是高度还是区位在当时校园内都具有绝对的优势，一跃成为令人瞩目的校园最高建筑，直到百年校庆时被综合楼超越。第二次改扩建为90年代，随着信息发展迅速，图书馆在原来书库的西面加建了新的书库。进入千禧年后，图书馆的使用再次暴露出种种问题——各种设施的老化、地基沉降、使用空间不足等，致使图书馆迎来了第三次改造。这次改造重心是大厅和塔楼，利用玻璃和钢结构在原有内院的位置上打造了一个与历史红砖墙脱开的椭圆形大厅，并沿椭圆形玻璃墙对称设置了两个弧形坡道，两个原本独立的塔楼在中央设置玻璃盒连接起来，并对每层的公共走廊和阅览室进行了分隔和规定，最大限度保留了极富历史特征的旧书库结构体系，将旧书库和西侧新书库之间搭建一个玻璃通廊，将西侧新书库改造为科技阅览室。

同济大学图书馆历经三次改造，成为校园

Experience from Guidance on Conservation and Renovation Design of Tongji University Library 2018

Zuo Yan, Lin Yi

The "Interior Design 6+" 2018 (Sixth Year) Joint Graduation Project Event, hosted by the IID-ASC, is themed on "National Quintessence Continuation—Historic Building Conservation and Reuse Design". Tongji University chooses the conversation and renovation of Tongji University Library, one of the buildings located in the main campus in Siping Road, Shanghai City that carry the most unforgettable memory of the campus and have the best historical look, as this year's graduation project theme. On the one hand, the students can have a basic cognition and attitude towards the conservation value of campus culture and historic buildings on campus; on the other hand, the students can identify main current problems after in-depth research, to have their integrative design ability and practical problem solving ability improved. Especially, in terms of lighting environment, they can develop improvement measures to have the project implemented from the "whole" to "parts".

1. Interpretation of the Historical Change and Value of the Library

The Tongji University Library in the main campus in Siping Road, Shanghai was founded in the 1960s, when Tongji University's gate, south and north buildings and Chairman Mao's statue, were built. The library is an important landscape building on the historical central axis of the campus. The library, which has a history of over 50 years, has been reconstructed or extended for three times. At first, the library was a two-storey building of masonry-concrete structure, and its facade is a red brick wall. Spatially, the building is shaped like two mouths superimposed together. This spatial pattern is related to the library's usage mode in those days, and the compact three-storey book storehouse is a non-open area, where books can be borrowed through librarians.

The first extension and reconstruction dates back to the 1980s. Due to the limited land, two cantilever towers were built in the original central inner garden to solve the problem of the lack of reading space. The tower structure is the reinforced-concrete core-tube cantilever structure that was fashionable in those days, and the cantilever length is nearly 10m. Structural innovation was of leading significance at that time, and the 11-storey tower facade is covered with mosaics, looking like red brick color. After extension, the building had overwhelming superiority in both height and location on campus at that time, thus quickly becoming the remarkable highest building on campus, but it was overtaken by the multiple-use building when the university celebrated its 100th birthday. The second extension and reconstruction dates back to the 1990s, when a new book storehouse was built to the west of the original one to cater to rapid information development. After 2000, the use of the library caused a number of problems: aging of facilities, foundation settlement and lack of space, which brought about a third renovation of the library. This time, the lobby and towers were mainly renovated. An elliptical lobby was built of glass and steel in the inner garden, and it is disjointed with the red brick wall. And, two arc ramps were set up symmetrically along the elliptical glass wall, connecting the two towers, which had been independent of each other, together through a glass box set up in the middle. The public corridor and reading room on each floor were partitioned and regulated, maximizing the preservation of the

内最具历史价值和集体记忆的建筑。此次学生的毕业设计是对该建筑的第四次改造，在设计改造前，充分了解、尊重和保护建筑遗存的风貌特征要素尤为必要，保护什么、怎么保护，这是开展功能定位和更新改造的第一步。校图历史上的三次改造代表着20世纪60年代、80年代及其千禧年三个不同时期同济图书馆的功能需求、建筑形态以及结构技术的历史演变，而结构改造上的大胆创新以及从红砖墙、马赛克到钢结构的材质转变可以折射出其背后社会、经济、文化的发展需求和时代特征。对于学生来说，既要了解和分析图书馆在建成50多年来自身功能不断发展沿革的内在规律和趋势，也要在新一轮更新改造中将校图历史特征要素周全巧妙地保留下来，找到新旧结合的适宜方式。这是该课题的最大难点和挑战。

2. 本科学习中最深入的设计调研

以往课程设计一般都会安排两周的基地调研和资料收集，但这次因为是毕业设计加上课题的高难度，基地调研和文献查阅成为学生了解基地现状、图书馆未来发展趋势以及针对现状问题寻求设计突破点的非常重要的环节，必须高度重视并留出较为宽松的时间安排。在老师们的启发下，学生们渐渐进入角色，以小组方式设计问卷调查表，开展人物访谈和实地深度调查，学生之间有分工合作，经过优化问卷质量、扩大访谈对象，最终访谈了图书馆内使用者100人，图书馆外南北教学楼使用者20人以及图书馆管理和工作人员。调研问题大概分为以下几类：服务类（包括对功能分区、环境、活动形式的建议等）、管理类（包括对门禁形式、座位使用模式、图书分类、查询和借还的建议等）和设施类（包括对桌椅设施、书架、灯光、导视系统、卫生间及其他服务区的建议等）。通过对调研数据的整理和分析，发现图书馆在现状使用上基本满足了大多数人的要求，但在一些细节方面如电源插座数量、书籍查阅、研讨空间等方面不尽如人意，存在阅览空间形式单一、设施设备落后、管理模式落后、空间浪费等问题，这些问题继而引发许多使用上不便利的情况。

调研分析发现空间方面问题最多的集中在图书馆大厅和旧书库。通过对大厅和书库部分的定点统计，整个大厅缺乏自由讨论交流区域，空间利用率非常低，实际上大厅的行为活动需求超过预期设想，平时少有人去的环形坡道不时会有讨论、接电话等行为发生；而三层旧书库的结构一体化构造被保留下来，但密集化的

old book storehouse structure system with outstanding historical features. A glass vestibule was built between the old book storehouse and the new book storehouse, transforming the new book storehouse into a science & technology reading room.

The Tongji University Library was renovated for three times, becoming the campus building with the greatest historical value and collective memory value. This time, the students' graduation project is a fourth-time renovation of the building. Before renovation design, it is particularly important to fully understand, respect and protect the scene and features of the building remains. What should be protected? How it should be protected? This should be done in advance of functional localization, renovation and transformation. The three times of renovation of the library represent the historical evolution of the functional requirements, architectural form and structure technology in three different periods, namely the 1960s, the 1980s and the 2000s. The bold innovation in structure reconstruction, as well as the change of materials from red brick to mosaic and to steel, can reflect the requirements of the social, economic and cultural development and the characteristics of the times. For the students, they should not only understand and analyze the inherent law and trend of the development and evolution of the library functions since its construction, but also well preserve all the historical features of the library, and find a proper way to combine the old with the new. This is the biggest challenge for the project.

2. The Most In-depth Design Research during Undergraduate Studies

In the curriculum offered in the past, two weeks would be spent on field research and data acquisition, but this time, field research and literature review became very important to the students, who had to understand the current situation of the site and the future development trend of the library, and make breakthroughs in design by solving practical problems while doing the high-challenging graduation project. Inspired by the teachers, the students gradually adjusted to the work and designed a questionnaire within the team to conduct interviews and in-depth field research. They divided the work, but cooperated with one another as well. By optimizing the questionnaire, they finally interviewed more people, including 100 users of the library, 20 users of the south and north teaching buildings, and the librarians. The questions in the questionnaire are roughly divided into the following categories: Service-related questions (including advice on functional partitioning, environment and event form); management-related questions (including advice on entrance guard form, seat usage mode, book classification, search, borrowing and return); facility-related questions (including advice on desks and chairs, bookshelves, lighting, the signage system, toilet and other service areas). The compilation and analysis of the research data show that the library can basically meet most people's needs, but some details, such as the number of electric outlets, book search and the discussion space, are unsatisfactory. Moreover, there are problems such as the single reading space form, backward facilities and management mode, and wasted space, which make it inconvenient to use the library.

钢结构体系使得书库一层、二层的高度偏低，柱子之间距离只有1米间隔，无法容纳两人并排穿行，且其所处位置在大厅中轴线上，书库的藏书功能日渐弱化，在千禧年的第三次改造后沦为图书馆去往西侧学苑食堂的必经通道，穿行过程中给人造成迷失和压抑的不适之感。

光环境改善是此次设计改造的一个亮点。在改造前对校图光环境现状情况要有一个基本认识。学生分组后选择图书馆中几个主要空间，运用亮度相机拍照，分析出图书馆在一天内的日光照度变化规律，发现不同使用区域的日光变化存在着较大差异。门厅附近，室内外照度差异最大，一进门就会有一种突然黑下来的感觉，而穿过门厅后进入大厅照度明显提高。裙房部分的阅读区域照度较高，虽然日光会随着空间的进深急剧递减，但人工光起到了良好的补光作用，不过有些地方灯具会有眩光。大厅内玻璃顶部大部分区域被塔楼底部所遮挡，尤其是中部直射日光少，通过太阳轨迹的模拟后发现夏天的某一时段大厅四角会有一段时间有直射光进入，书库缺乏自然光，由于书库层高低，照度明显没有达到标准，垂直照度普遍没有满足视看要求，整个书库氛围昏暗压抑。科技阅览厅照度偏高，且眩光严重，书库与科技阅览厅之间的玻璃通廊为玻璃顶，阳光充足，同时热辐射也非常高，虽然顶部加装了格栅，遮阳作用并不明显。塔楼光照环境差别很大，南塔楼每天日照时间长，阳光充足，但建筑立面缺乏遮阳考虑，整个室内环境仅靠薄型窗帘遮光，会影响到内部空间的阅读使用。书架部分由于顶上缺少照明，垂直照度递减严重，电梯等候室和应急通道照明也并非理想，亟待改善。

除了对图书馆建筑的本体调研外，学生还走访调研了如校史馆、校博物馆、院史馆、图书分馆、档案馆等学校其他文化场所，发现除了图书馆，校史馆、校博物馆也扮演着重要角色，但是校博物馆和校史馆的地理位置较为偏僻，很难吸引大量人流，而各个院系的院史馆和图书馆也分散校园各处，其中大多数资料也不对外开放，周末都处于关闭状态，因此校图书馆成为整个校园最便利、最重要的阅览场所了。此外学生还调研了一些特色书店，像钟书阁、志达书店、衡山和集等，还组织参观了一次复旦枫林校区的图书馆。三个多星期的调研和文献检索确保学生了解了该课题的复杂性和真实性特征。

The research and analysis show that most space problems exist in the library lobby and old book storehouse. Fixed-point statistics of the lobby and book storehouse indicate that there is a lack of a free discussion area in the lobby, and the space utilization ratio is very low. Actually, the demand for behavior activities in the lobby is higher than expected, and on the annular ramp, where there are usually few people, there are some people discussing with each other or answering the phone from time to time; the integrated structure of the three-storey old book storehouse is retained, but the dense steel structure system makes the first and second floors a little low, and the inter-column space, which is only 1m wide, cannot accommodate two people walking side by side. Moreover, the book storehouse is located on the central axis of the lobby, and its storage function has been gradually weakened. After the third renovation in the 2000s, the book storehouse was reduced to a passage which must be passed to the dinning hall to the west of the library. When walking through it, people feel as if they were lost, and are depressed.

The improvement of the lighting environment is a bright spot in the renovation design. Before renovation, the students would better have a basic knowledge of the lighting environment in the library. After divided into 2 teams, the students chose several main spaces in the library and took photos of them with a luminance camera. An analysis of the photos reveals the intraday change law of the solar illumination in the library, and shows that there is a great difference in daylight change from area to area. There is the greatest difference in indoor and outdoor illumination near the entrance hall, and it would turn dark suddenly when one walks through the door, but the illumination begins to increase obviously as one moves inwards. There is high illumination in the reading area in the skirt building. Although the solar illumination weakens sharply in the depth of the space, artificial light works well there, only that there is dizzy light in some areas. Most of the glass ceiling in the lobby is sheltered from the sun by the bottom of the tower building. Especially, there is little direct sunlight in the middle.

A simulation of the sun trajectory reveals that the sun shines directly on the four corners of the lobby for some time in a certain period of time in summer, but the book storehouse lacks natural light. Since the ceiling of the book storehouse is low, the illumination cannot meet the standard, neither can the vertical illumination make it bright enough to read books, so the whole book storehouse is dimly lit and dreary. The illumination in the science and technology reading room is a little too high and severely glaring. The glass vestibule between the book storehouse and the science and technology reading room is equipped with a glass ceiling, in which ample sunlight comes. Meanwhile, thermal radiation is very high, and even though there are georgics on the roof, the sun-shading effect is not so obvious. There is a big difference in illumination between the two tower buildings. The south tower is in the sun for a long time every time and receives sufficient sunlight, but the building elevation doesn't have a good shading effect, so the whole indoor environment has to be protected against the sun by thin curtains, which have an adverse impact on reading indoors. For the bookshelves, vertical illumination weakens progressively sharply due to the lack of lighting on the top. Besides,

3. 光环境改造的设计指导

根据任务书要求，学生从设计之初就把光环境作为研究对象进行专项调研和评估，通过切实地体验光、感知光的过程，结合相关技术支持进行光环境的实测，在感性认识和理性评估两个层面对光环境进行主动探索，进而原发性地将光环境需求作为设计的出发点之一。

在具体的设计过程中，学生被要求从视觉作业的需求以及心理舒适度的需求两个角度设定光环境设计目标，并从自然光与人工照明两条路径展开设计。由于本科阶段培养体系对于光环境的教学总体上仍显薄弱，在毕设过程中穿插必要的光环境专业理论知识的补充和梳理、强化教学环节，并辅以相关新技术的介绍和技能的快速培训，如光环境计算模拟软件的使用。此外，实物模型的推演也是设计中的关键环节之一。由于自然光在空间的分布仅与空间的比例关系，因此实物模型能够有效、快速地模拟采光效果，并提供最为直观、准确的光照分布效果。对于不可确定又难以计算模拟的部分，学生会采用实物模型进行设计推演。此时的模型不同于最终的效果模型，不追求精致美观，而是服务于对室内光环境的观察分析。如毛燕同学针对现有书库部分一、二层空间狭小、封闭、缺乏自然光的情况，提出把现有三层屋顶导光管改造为贯通三层的光井的设想。但是如何在有限的空间比例中把光有效地引导进入各层空间成为设计的难点。在其对光与材料的研究中发现光线可以在透明亚克力杆件内部传导，并在管壁和端头散射出。于是她制作1:50局部光井模型，用亚克力材料模拟棱镜导光管的效果进行模拟实验，评估光照效果，最终决定采用长短不一棱镜导光杆件将室外自然光按需要传导到不同的楼层中的设计方案。这一过程也使得方案具备了一定程度上的可行性和有效性。

总之，光环境作为室内环境不可分割的一部分无法孤立进行分析和设计，应结合整个室内外空间整体的功能定位和改造来完成。

作者单位：同济大学建筑与城市规划学院

the elevator room and escape trunk are not in good illumination conditions either, with illumination urgently needing to be improved.

3. Guidance on the Renovation Design of the Lighting Environment

According to the requirements of the assignment, the students should focus on the lighting environment from start to finish and do research on it. After experiencing the lighting and perceiving its process in the field, the students detected the lighting environment with related technology. Then by perceptual knowledge and rational evaluation, they made an active exploration on the lighting environment, and then primarily took the demand of lighting environment as a design starting point.

In the course of design, the students were asked to set a goal of lighting environment design from the perspective of visual demand for reading and psychological demand for comfort. Since lighting environment teaching is not an important point in the undergraduate educational system, the students were taught the professional theoretical knowledge of lighting environment intentionally during the graduation project event. Moreover, they were taught related new techniques and skills through short-term training, such as the use of lighting environment calculation and simulation software. In addition, physical model deduction is also critical to the design. Since the spatial distribution of natural light is only related to the space scale, a physical model can effectively and quickly simulate the daylighting effect, and provides the most intuitive, accurate effect of illumination distribution. For what cannot be determined or is hard to calculate and simulate, the students would use a physical model for design and deduction. This model, different from the final effect model, needn't be beautiful, but just used to observe and analyze the indoor lighting environment. For example, of the students, Mao Yan suggested transforming the light pip on the three-storey roof into a light well that runs through the three floors to solve the problems in the first and second floors, such as narrow space, closed space, and lack of natural light. However, it is difficult to effectively import light into all the floors in spite of the limited space scale. While researching lighting and materials, she found that light could travel through a transparent acrylic bar, and scatter through the pipe wall and end. So, she made a 1: 50 local light well model and simulated a prism light guide with an acrylic material to evaluate the lighting effect. Finally, she decided to transmit natural light into different floors with prism light guides of various lengths. The simulation experiment makes her design proposal feasible and effective to some extent.

In a word, the lighting environment, as part of the indoor environment, cannot be analyzed or designed in isolation, but should be renovated in accordance with the functional localization of the indoor and outdoor space as a whole.

建筑学专业毕业设计成果评价模式的探讨

周立军 马辉 兆晕

建筑学专业的毕业设计是建筑学本科生五年建筑设计系列课程的收官课程。毕业设计过程历时半年，是学生专业综合能力的集中体现。作为检验毕业设计教学成效的标尺，评价制度研究是至关重要。根据人才培养目标和理念，建立科学的评价标准。是激励学生乐观向上、自主自立、努力成才的保证。因此，在传统的建筑设计评图方式基础上，笔者提出了形成性评价、开放性评价、综合性评价相结合的评价理念。

1. 形成性评价

过程评价（形成性评价）是1967年由美国哈佛大学斯克里芬（M. Scriven）在开发课程研究中提出的。布卢姆（B. S. Bloom）将其引入教学领域，提出了掌握学习的教学策略，取得了显著的成绩。布卢姆侧重对学习过程的评价，并把评价作为学习过程的一部分。布卢姆主张教学中应更多地使用另一种评价方法——形成性评价或形成性测验。布卢姆指出，评价与评分不是同一个意义。不评分也能够对学习结果作出评价，形成性评价就是一个例子。形成性评价的有效的程序是：把一门课分成若干学习单元，再把每一单元分解成若干要素，使学习的各种要素形成一个学习任务的层次，确定相应的教育目标系统；形成性测验常常被用来为学生的学习定速度，保证学生在从事下一个学习任务之前，完全掌握这一单元的内容。

形成性评价是对学生的学习过程进行的评价，旨在确认学生的潜力，改进和发展学生的学习。形成性评价的任务是对学生日常学习过程中的表现、所取得的成绩以及所反映出的情感、态度、策略等方面的发展做出评价。其目的是激励学生学习，帮助学生有效调控自己的学习过程，使学生获得成就感，增强自信心，培养合作精神。形成性评价不单纯从评价者的需要出发，而更注重从被评价者的需要出发，重视学习的过程，重视学生在学习中的体验；强调人与人之间的相互作用，强调评价中多种因素的交互作用，重视师生交流。

毕业设计课程评价标准针对过去评价只重视总结性评价而产生的种种弊端，提出总结性评价与形成性评价相结合，更关注形成性评价的新理念，形成从开题、中检、验收、答辩等一系列的过程式累加的评价体系。

2. 开放性评价

开放式评价是指评价主体的多元化、评价形式的公开性和多样化。它主要是包括学校、教师、学生之间对学生在知识与技能、过程与方法以及情感、态度、价值观等方面发展状况

The Ways to evaluate the Results of Architecture Graduation Project

Zhou Lijun, Ma Hui, Zhao Hui

The graduation project of architecture is the last of the five-year undergraduate architectural design courses. Graduation project, which takes half a year to do, is a concentrated reflection of students' professional comprehensive ability. It is of crucial importance to research the evaluation system because it can be used to test the outcomes of graduation project teaching. Scientific evaluation criteria should be set in accordance with the talent training objective and idea to encourage students to be optimistic and cooperative, autonomous and independent, and hard-working to become a talented person. Therefore, on the basis of the traditional architectural design review mode, we propose combining formative evaluation, open evaluation and comprehensive evaluation together.

1. Formative Evaluation

Process evaluation was advance by M. Scriven, a professor at Harvard University, in 1967 in the course of curriculum development. B. S. Bloom introduced it into the teaching domain and advanced a teaching strategy named mastery learning, achieving remarkable results. Bloom focused on evaluating the learning process and considered evaluation to be part of the learning process. He suggested making more use of another evaluation method— formative evaluation or formative test—in teaching. According to him, evaluation and scoring have different meanings. Learning outcomes can be evaluated without scoring, and formative evaluation is one example. Formative evaluation has the following effective procedures: divide 1 course into a number of learning units, and then divide each unit into a number of elements to form a learning task system and build an educational target system; formative test is often used to set a speed for students' learning, to ensure that they will fully understand the content of one unit before fulfilling the next learning task.

Formative evaluation is an evaluation of students' learning process, aimed at conforming students' potential and improving their learning. Formative evaluation is carried out to appraise students' daily learning performance, achievements, as well as their feelings, attitudes and learning strategies. Its aim is to encourage students to study and help them effectively control their learning process, so that they could develop a sense of accomplishment, increases self-confidence and cultivate the team spirit. Formative evaluation does not only start from evaluators' needs, but more from the needs of the evaluated. It values the learning process and students' learning experience; stresses inter-people interaction, inter-factor interaction and teacher-student interaction.

Compared with the conventional evaluation method that focuses only on summative evaluation and therefore has many defects, the evaluation criteria for graduation project curricula propose combining summative evaluation with formative evaluation, with more focus put on the new idea of formative evaluation, forming an evaluation system containing opening report, in-process inspection, acceptance inspection and oral defense.

2. Open Evaluation

Open evaluation refers to the diversification of evaluation subjects and the openness and diversification of evaluation forms. It basically consists of the university's, teacher' and students'

的评价。教学评价不单是在学校中进行，而应在整个社会中进行。它应该在相应的情景中进行，应以专题作业、过程作品集等形式架起校内活动和校外活动之间的桥梁。评价方案要考虑到个体之间的差异、发展的不同阶段和专业知识的多样化。评价特别强调对个体智能强项的识别，充分发挥学生个体的潜能。评价的过程应该是贯穿在整个学习过程之中，无论是课堂上的学习活动，还是学生参加设计等实践活动等，均含在评价之内，并注重大众评价与专业评价相结合。如展览式评价就是评价主体从毕业设计组教师进一步扩展到全体师生乃至社会大众。这种评价模式由于主体更为开放，学生作品直接与潜在使用者接触，使学生对自己作品的社会接受度有更深的理解，对专业更有成就感，其社会影响远远超过了专业评价层面。

3. 综合性评价

综合式评价一方面是指学生毕业设计成果的综合性，如作业和图纸，不仅展示作品的设计和设计逻辑，还有表达建筑的空间、功能和技术理念，审查设计图纸的规范性和完成度，同时，学生还要在设计过程中通过图纸、模型和多媒体与教师进行汇报和交流，培养自身的表达能力。另一方面是指教师评委的综合性，。综合式评价模式融合了专业评委与相关专业评委、指导教师与非指导教师、校内评委与校外评委以及国外和企业评委等，突出"海内外结合、校企结合、本科教学与实践项目相结合、相关专业结合"等特色。综合性评价方式的开展上还可以继续探讨新型的评价方式，如今是网络时代，评价方式可以借助多媒体的相关工作开展。同时目前课程评价一直限于本专业，也可以去试图通过网络投票的方式进行公众评价，在整个学校的师生之间开展，甚至有些课程设计成果还可以在社会上开展相关评价。综合式评价使建筑学毕业设计作品的评价体系更为科学系统，对类似作业的评价也具有一定的借鉴意义。

4. 结论

德国教育家第斯多惠说过：教学的艺术不在于传授本领，而在于激励、唤醒和鼓舞。这句话充分说明了对学生评价的作用，评价也是一门艺术。评价学生所起的是一种激励与促进作用，让不同潜质的学生能通过评价看到自己的优点，培养自己的成就感；同时培养一种团结向上的协作精神，同学之间只有在协作的氛围中才能获得小组评价的好成绩；更重要的是培养学生对建筑设计的兴趣，是一种快乐与渴望中学习，渴望与同学之间的竞争与合作。因

evaluations on students' development in knowledge, skills, process, method, emotion, attitude and values. Teaching evaluation shouldn't be made in universities alone, while it should be made in the society as well. It should be made in a related situation and build a bridge between intramural activities and extramural activities in the form of special topic assignment and process work portfolio. The evaluation program should give consideration to individual differences, different development stages and the diversification of professional knowledge. Evaluation lays a special emphasis on the identification of individuals' intelligent forte, in order to fully tap individual potential. Evaluation should run through the whole learning process, including learning activities in class and practical design activities. Also, public evaluation is combined with professional evaluation. Spectacular evaluation means that the evaluation subjects are no longer the graduation project supervisors alone, but also all the teachers and students and even the public. In this evaluation mode, there are more subjects, and students' works thus may be seen by potential users, so students can get a deeper understanding of the social acceptability of their works, and develop a stronger sense of accomplishment. So, the social influence of this evaluation mode is far greater than professional evaluation.

3. Comprehensive Evaluation

Comprehensive evaluation, on the one hand, refers to the comprehensiveness of graduation project achievements. Taking assignment and drawing for example, they not only display design and design logic, but also present the spatial, functional and technical idea of architecture, and review the normativity and completeness of the design drawing. Meanwhile, in the process of design, students need to report to the teachers and communicate with through the drawing, model and multimedia, to enhance their own expressiveness. On the other hand, it refers to the comprehensiveness of evaluation subjects. The comprehensive evaluation is usually made by professional judges, associate professional judges, supervisors, non-supervisors, intramural judges, extramural judges, foreign judges and enterprise judges. That highlights the features of comprehensive evaluation, such as "domestic-foreign combination, university-enterprise combination, combination of undergraduate teaching with practical projects, and combination of related specialties". For the implementation of comprehensive evaluation, new modes can be developed. In the present era of Internet, evaluations can be made through multimedia. Currently, curriculum evaluation is always restricted to the specialty itself, but we can try to carry out public evaluation among all teachers and students through Internet voting. Even some curriculum design results can be evaluated socially. Comprehensive evaluation makes the evaluation system of architecture graduation project works more scientific and systematic, and is of some reference significance to the evaluation of similar assignments.

4. Conclusions

German educator Diesterweg said, "The art of teaching lies not in knowledge instruction, but in motivation, arousing and inspiration." This fully demonstrates that evaluation of students is also an art. Evaluated, students are so motivated that they, despite different potential, can see their own strengths and develop a sense

此,建筑学毕业设计的评价可以是多方位、多角度、多层次、多元化的,教师和学生可将建筑设计的学习评价延伸置课堂内外,以此来不断地激励学生的进步。

作者单位:哈尔滨工业大学建筑学院

参考文献:
[1][德]第斯多惠.德国教师培养指南[M].袁一安,译.北京:人民教育出版社,2001:72-74.
[2]Richard I.Stiggins.促进学生的学习参与式课堂评价[M].国家基础教育课程改革"促进教师发展与学生成长的评价研究"项目组,译.北京:中国轻工业出版社,2005:155-185.
[3]张咏梅,孟庆茂.新课程下的学业成就评定:观念与方法[J].全球教育展望,2003(11):34-37.
[4]DeciE.L.,VallerandR.J.,PelletierL.D.,etal.Motivation and Education: The Self-determination Perspective [J]. Educational Psychologist, 1991(26): 325.

of accomplishment; moreover, they have a good sense of team spirit and realize that only in a collaborative atmosphere can they be well evaluated; what's more, students can develop an interest in architectural design, feel happy to study and dream of competing and cooperating with their classmates. Therefore, architecture graduation project can be evaluated multi-dimensionally, from multiple angles, at multiple levels and diversely. Teachers can teach and evaluate architectural design inside and outside the classroom, to constantly encourage students to make progress.

Author Affiliation: School of Architecture, Harbin Institute of Technology

环境设计专业本科毕业设计联合指导模式探索

刘晓军　王敏

教学目标是教学的出发点和归宿，它从根本上支配并指导着教学实践活动的各个环节。在环境设计学科的本科教育阶段，现时代的市场需求及实践应用能力的培养应该被引起重视，培养学生以行业需求为导向的设计思维，而联合毕业设计活动正是给我们提供了这样一个广阔的交流平台。

一、活动特点简述

"室内设计6+"联合毕业设计活动是一个联合全国多所高校、结合设计企业实际项目的综合毕业设计活动。活动至今已连续成功举办六届，每届都由参加活动的高校联合当地设计单位出题，针对当下的社会热点问题，从学术探讨的角度出发，尝试用设计来解决社会现实问题，具有校企联合、多校联合、导师联合等特点，选题具有针对性，对环艺本科学生的教学具有鲜明的专业导向性。

1. 校企联合、互利共赢

一方面，校企联合教育是培养应用型人才的一个有效途径，企业的介入打破了传统的教学成果评定标准。学生的设计作品除了有良好的设计理念、设计方法及设计表现外，还要顺应社会发展要求，体现时代精神，促使高校更新教学内容、教学环节、教学方法及手段，从而满足经济和社会的发展需求。

另一方面，企业参与高校的毕业设计全过程，通过学生各种发散的、原创的、新奇的设计创意，激发自身设计团队的创作灵感。高校是人才的聚集地，通过校企合作关系的建立，让企业更多的了解学校教育及优秀学生，为学生的就业做好铺垫，从而实现资源共享、互利共赢、共同进步。

2. 多校联合、取长补短

"室内设计6+"联合毕业设计活动旨在加强室内设计师培养的针对性，促进高校之间在室内设计学科建设和教育教学方面的交流，继而实现以行业需求为导向的专业学科发展目标。每所高校都有其自身的教学特点，在此过程中，各个高校之间通过对学科建设、培养目标、教学手段及方法、学术研究等方面的探讨和交流，互相取长补短，共同推进环艺专业学科的发展建设。

3. 导师联合、全面指导

联合毕业设计活动具有理论性和实践性的双重特色，高校导师和企业导师，对学生的毕业设计作品分别给予学术性和落地性的指导意见。此外，来自各个高校的不同导师，在大学科的统一下各自近期的研究方向也略有不同。

An Exploration of Modes for Joint Guidance on Undergraduate Graduation Project of Environmental Design

LIU Xiaojun, WANG Min

A teaching objective, which is the starting point and destination of teaching, fundamentally dominates and guides the whole process of teaching practice. At the undergraduate education phase of environment design, contemporary market demand and practical ability training should be brought to the forefront, to develop students' industry demand-oriented design thinking. The joint graduation project event provides us with a wide exchange platform of this kind.

I. About Event Characteristics

The "interior Design 6+" Joint Graduation Project Event is a comprehensive graduation project event joined by a number of national universities with practical project topics assigned by design companies. The event has been held for six times so far. For joint graduation project teaching, every year's topic is a social hotspot, which is finalized by the university itself with the design company on a realistic basis. From an academic perspective, design is used to solve practical social issues on the basis of university-enterprise cooperation, inter-university cooperation and inter-supervisor cooperation. The topic is issue-relevant and of guiding significance to the teaching of environment art design.

1. Win-win University-enterprise Cooperation

On the one hand, university-enterprise cooperative education is an effective way to raise applied talents, and an enterprise's intervention breaks the traditional evaluation standard of teaching achievements. The students' works should contain a good design idea, method and performance. Besides, the works should meet the requirements of social development, reflect the spirit of the times, and propel the universities to renew teaching content, process, methods and means to meet the requirements of economic and social development.

On the other hand, in the whole process of graduation project design, the corporate participants help students to think divergently and make original and novel designs to inspire them. A university is a place where talented people gather and live together, so through a university-enterprise cooperative relationship, enterprises can better understand university education and excellent students, and pave the way for student employment, so as to realize resource sharing, achieve a win-win situation and make common progress.

2. Inter-university Cooperation to complement each other's advantages

The "interior Design 6+" Joint Graduation Project Event aims to enhance the pertinence of interior designer cultivation, promote inter-university exchanges of ideas on interior design discipline construction, education and teaching, and then achieve specialized discipline development to meet industry needs. Every university has its own teaching characteristics. In this process, the universities learn from each other's strong points and close the gap by discussing on discipline construction, training objectives, teaching means and methods, and academic research, to jointly promote the development and construction of environment art design.

3. Supervisor Cooperation to offer Comprehensive Guidance

The joint graduation project event is a theoretical and practical activity, in which university and enterprise supervisors give academic

各个导师对学生设计作品的解读及指导意见，切入角度不同，从而使得同一个问题立体化、多样化，学生可以得到来自企业和不同高校导师的多方位指导，使得对其设计项目的理解更加全面而深入。

二、环境设计专业特色

1. 多学科、综合性

环境设计专业是一个跨学科的综合性专业，涉及美术、雕塑、城市规划、建筑学、风景园林、植物学、生态学、室内设计、环境物理、人体工程学、家具设计、装饰艺术、材料学、心理学等众多学科，具有多学科交叉渗透的特点。基于这一专业特点，在人才培养方面，既要注重培养学生理论知识体系的全面性，又要注重各学科之间综合应用能力的培养。

2. 实践性、应用性

环境设计学是艺术设计学科的一个分支体系，它区别于纯艺术专业，具有艺术与技术相结合的特点，集功能、艺术与技术于一体。在教学过程中，既要激发学生的创造性思维意识，又要传授多样化的技术实施手段，引导学生从功能出发、以人为本来进行设计构思。

3. 时代感、适应性

环境设计不是一成不变的，不同历史时期的设计作品都有其自身的时代特色，要顺应时代的发展和社会的进步需求，在尊重历史文脉的同时突出时代感，满足不同时代的人对环境的多方面需求。

三、校企联合教育的优越性与必要性

1. 产学研一体化

对于现时代高速发展的现状，只有将科研、教育、生产有机的结合，才能更有力的推动社会经济的发展。联合毕业设计活动有力的贯彻执行了这一方针政策，通过该活动，让学生深切体会到自己的设计作品在转化成实际成果过程中的不足之处，以此发现问题，改正问题，进而方向明确的引导日后的设计和研究。通过这样的合作交流，高校在教学及科研方面也会针对现状，适时作出相应调整，同时这也是提升企业和产业竞争力的有效方式。

2. 使设计目的更加明确

企业是人才培养的风向标，在企业的参与及建议指导下，学生的设计目的更加明确，实践导向性更加直观。设计不是为了"完成作业"，引导学生带着设计师的使命感去发现问题，用设计来解决现实问题，并对设计方案进行可行性分析，更加深刻的植入整个设计过程及细节

and pragmatic guidance on students' graduation projects. Besides, the supervisors from different universities now have different research directions under the same discipline. There being a difference in interpretation and guidance on students' design from supervisor to supervisor, the same issue becomes stereoscopic and heterogeneous. Students can get different guidance from the enterprise and university supervisors, so as to get a more comprehensive and profound understanding of design projects.

II. Features of Environmental Design

1. Multidisciplinary and Comprehensive

Environmental design is an interdisciplinary comprehensive specialty, covering fine arts, sculpture, urban planning, architecture, landscape architecture, botany, ecology, interior design, environmental physics, human engineering, furniture design, decorative art, material science and psychology. Considering the features of specialty, in terms of talent cultivation, we should not only teach students comprehensive theoretical knowledge, but also develop their ability to apply interdisciplinary knowledge.

2. Practical and Applicable

Environmental design is a branch of art design that, different from fine art, is a combination of art with technology, featured by the fusion of function, art and technology. In the process of teaching, we should not only encourage students to think creatively, but teach them diverse technological application means, to instruct them in design conception from a functional perspective.

3. Temporal and Adaptive

Environmental design is not unchanging, while design works of different historical periods have their own time characteristics. So, the works should conform to the requirements of historical development and social progress, and highlight the temporality while respecting the historical context, to meet different generations' diverse needs for the environment.

III. The Superiority and Necessity of University-Enterprise Joint Education

1. Industry-university-research Integration

In the present era of rapid development, socioeconomic development won't be promoted unless scientific research, education and production are organically combined together. This policy is effectively implemented at the joint graduation project event, where the students clearly realized the deficiency of their work in terms of achievements transformation, and thus identified and solve problems, finding a clear direction for future design and research. After cooperation and exchanges this way, the universities will adjust the content of teaching and scientific research to some extent. Besides, enterprise and industry competitiveness can be enhanced effectively.

2. A Clearer Design Goal

The enterprise is the weather vane of talent cultivation, so under the guidance of the enterprise, students have a clearer design objective, which is more practical. Design does not aim to "finish assignments", but guide students to identify problems with a sense of mission as a designer, then solve the problems by design, and make a feasibility analysis of the design scheme, with more profound practice-oriented thinking consciousness implanted into the whole

都要以实践为导向性的思维意识。

3. 使设计程序更加完整化

高校的设计教学大多止步于方案设计阶段，有了企业的介入，使学生真题真做，将调研、设计、实施等环节综合进行考虑，有效的解决了学生方案天马行空的普遍现象。校企协助共同推进，加强学生的实践思维，学会合理的协调设计创意与实际落地之间的矛盾，对于环艺专业的教育教学工作，具有深远的意义。

四、教学方法及教学手段的革新

基于环境设计专业的特点及校企联合教育的优越性，针对于环艺专业本科教学方法及手段，可以进行以下几点革新。

1. 以满足市场和社会需求为导向制定培养方案

围绕目前环境艺术设计产业的发展趋势，调整专业课程设置与教学内容，在保证学生掌握系统的专业知识、专业理论的基础上，适当增设部分专业应用型知识，有针对性的制定培养目标，注重学生个人能力的培养。环境设计专业学科具有多学科综合性的特点，学生就业面较宽广，因此如何将跨学科、交叉性知识合理导入到日常的教学工作中，使学生具有开放型的知识结构，同时防止理论知识相对于市场发展的脱节与滞后，是我们值得探讨和研究的内容。

2. 注重前期设计调研

设计调研除了记录现状以外，更重要的是感知思考、发现问题、提出问题，并在后续的设计当中解决问题，好的设计作品除了观赏性，其内涵在于用设计来解决实际问题。因此要引导学生进行有针对性并且深入的调查研究，调研的数据资料收集要完整，调研手段多样化，思路清晰目的明确，带着问题去调查研究，并从中找到答案，这样才能从根本上防止制式化的设计模仿，做出具有生命力的设计作品。

3. 实地参观考察，增设实践性课程

环境设计是一门集艺术、实践和市场为一体的综合性学科，传统的课堂教学模式，教学手段单一，教学资源有限，在知识传授的完整性和直观性上有客观的缺陷。因此，应该增设实践性课程，带学生实地参观考察，指导学生在真实的设计背景环境下灵活运用所学的理论知识，将所学、所想、所做有机统筹，使学生深刻理解"以理论知识指导设计实践，通过设计实践不断修正理论知识"的辩证设计观。此外，实践性课程的开展，以团队的方式共同完成设定项目，培养学生的团队合作意识，在团

design process and details.

3. A More Complete Design Process

University design teaching usually involves nothing more beyond conceptual design, but the enterprise can help students make designs in practice. Comprehensive consideration of research, design and implementation can effectively prevent students from make designs in an unconstrained style. Instead, the university and enterprise jointly strengthen students' practical thinking, teach them to rationally coordinate the conflict between design creativity and implementation. This is of far-reaching significance to the education and teaching of environment art design.

IV. Reform of Teaching Methods and Means

The teaching methods and means of undergraduate environment art design can be reformed as follows based on the features of environment art design and the superiority of university-enterprise joint education.

1. Training Program Development based on Market and Social Demand

Specialized curriculum setup and teaching content can be adjusted around the development trend of the environmental art design industry, and some professional applied knowledge can be added on the basis of enabling students to acquire comprehensive professional knowledge and theories to make a specific training objective that focuses on fostering students' personal competence. Environment art design is an interdisciplinary major, which involves a range of jobs, so we should discuss and research how to reasonably import interdisciplinary knowledge into day-to-day teaching to help student build an open knowledge structure. But meanwhile, theoretical knowledge must be prevented from being out of line with market development.

2. Laying Stress on Preliminary Design Research

During design research, students should not only record the present conditions, but also think with perception to identity problems, raise questions, and solve problems in follow-up design. Not only does a good design work have a ornamental value, but it has a connotation that practical problems are solved by design. Therefore, students should make a specific and in-depth research by diverse means with a clear objective under guidance and collect complete data. During research, they ought to look for solutions to problems, and only in this way can mechanical design imitation be completely eradicated and works with vitality be created.

3. On-the-spot Investigation and Practical Curriculum Setup

Environmental design is a comprehensive discipline that covers art, practice and market, but for the traditional classroom teaching mode, there is only a single teaching means and limited teaching resources, so the knowledge imparted is not integrated or intuitive. Therefore, practical curricula should be offered, and students can make an on-the-spot investigation and make a design in a real design environment based on what they have learned and thought by making flexible use of their theoretical knowledge, so as to get a profound understanding of the dialectical design concept that "design practice is guided by theoretical knowledge and theoretical knowledge is constantly corrected through design practice". In addition, after

队中发现并发挥自身优势、取长补短、互助合作，使学生在日后众多的职业选择中有的放矢。

4. 对学生思维意识的培养

设计思维决定其设计价值，本科教育与职业培训的最大区别就在于思维意识的培养。鼓励学生跨学科、多角度的思考设计问题，提升个人修养，从哲学辩证法及社会学等更加宽广的领域出发，探索人与自然、人与人、人与物之间的关系，将感性思维与理想思维相结合，提升设计作品的深度与内涵。

五、小结

自"室内设计6+"联合毕业设计活动开展以来，西安建筑科技大学艺术学院师生已连续参加六届，这是一个非常宝贵的学习与交流的机会，对我校环境设计专业全面提高教学水平，积极探索顺应市场发展需求的人才培养新模式，更新现有的教学手段及教学方法，提高学生的专业技能及综合素质等方面都具有深远的影响。

校企联合教育，高校和企业之间展开有效的合作，高校与高校之间进行深入的探讨和交流，通过企业的窗口开拓学生的设计思维，完善知识体系，增强专业技能，是开展实践教学模式的一个有效途径。以社会发展和市场需求为导向完善培养目标，在新形势下合理的更新教学模式，从而改变教育对市场的滞后性现状，提升高等院校人才培养的社会适应性。

作者单位：西安建筑科技大学艺术学院

practical curricula are offered, a specific project can be completed by a team to develop students' teamwork awareness, identify and use their own advantages in the team to work together by learning from each other's strong points and close the gap, so that they could choose a right job in future.

4. Cultivation of Students' Thinking Consciousness

Design thinking determines the value of design. The biggest difference between undergraduate education and vocational training is the cultivation of thinking consciousness. Students should be encouraged to think about design problems from multiple interdisciplinary perspectives, cultivate themselves, explore the relationship between man and nature, man and man, and man and things from philosophical dialectical and sociological perspectives, and combine perceptual thinking with rational thinking to boost the profoundness and connotation of design.

Since the founding of "Interior Design 6+" Joint Graduation Project Event, the teachers and students at School of Art, Xi'an University of Architecture and Technology have attended the event for 6 consecutive years. This is a valuable opportunity for study and exchanges, and of far-reaching significance to comprehensively increase our teaching level, actively develop a new mode of talent training in line with market development, renew the existing teaching means and methods, and improve students' professional skills and comprehensive qualities.

University-enterprise joint education refers to effective cooperation between universities and enterprises, as well as universities and universities, which make in-depth discussions and exchanges of views together. The enterprise is an effective to carry out practical teaching, because it can open up students' design thinking, which enables students to improve their knowledge system and enhance their professional skills. Social development and market demand help set a better training objective and renew the teaching mode properly under the new situation, to synchronize education with market, so that universities' talent training mode could better adapt to the society.

恭王府博物馆文化空间探索

杨琳 陈静勇 李奕慧

历经200多年风雨的恭王府是现存唯一保存完整的清王府，承载着浓厚的历史文化信息。具有历史、艺术、文化和社会价值，是王府类建筑的典范。恭王府经历了权倾朝野的大学士和珅、乾隆爱女和孝公主、嘉庆爱弟庆亲王永璘、同治光绪年间执掌大权的恭亲王奕䜣、意图恢复清朝统治的溥伟、有"南张北溥"之称的著名画家溥儒都曾在府中居住。目前恭王府是全国重点文物保护单位、国家AAAAA级旅游景区、国家非物质文化遗产展示与保护中心、国家一级博物馆，承担着旅游开放、博物馆、非遗展示与保护的职责，是活态文化空间、传统文化艺术展示平台、传承的文化场所。

1. 保护原则

1982年实施了《中华人民共和国文物保护法》，加强对于文物的保护，到目前为止《文物保护法》经过了五次修正，逐渐完善了文物保护法律体系。恭王府建筑的保护遵循《文物保护法》，按照真实性和完整性的保护原则，保存它们的全部历史信息。2002年颁布的《中国文物古迹保护准则》是对《文物保护法》深化的专业性文件。参照《威尼斯宪章》，结合中国文物保护长期的经验积累，促进中国文化遗产保护的发展。

文物建筑及历史地段保护的国际原则《威尼斯宪章》强调保护文物建筑的真实性，以结合科学技术手段修复文物建筑，提到对绘画、雕刻、装饰不得新建、拆除或变动，仅在因保护需要时可以取下的保护方法，再现历史文物建筑的审美和价值。《文物保护法》规定对历史文物建筑的修缮、保养、迁移以及使用时遵循不改变文物原状的原则。

恭王府有着浓厚的历史价值、美学价值等，在保护修缮问题上要积极贯彻各项历史文物建筑保护的法律条例。恭王府因经历多任府主的变更，每个时期的历史文化信息均在建筑装饰上得以体现。修缮时虽定位于同治、光绪时期的历史面貌，但对于府邸中留存的同光时期以前的装饰也加以保留，展现不同时期的装饰特点。

"有历史根据的、按历史根据进行修复；找不到历史根据的、充分听取专家意见，按专家指导进行修复；无历史根据、专家也拿不准的、按现状进行修复"是2005年恭王府修缮时确立的修缮原则，这与《威尼斯宪章》和《文物保护法》中对历史文物建筑保护的原则相一致。对原构件尽可能保留利用，展示清王府历史风貌。不能利用的构件进行原状陈列，保留历史信息。对于传统装饰构件要以传统营造技

An Exploration on Prince Kung's Palace Museum

Yang Lin, Chen Jingyong, Li Yihui

The Prince Kung's Palace, built over 200 years ago, is the only well-preserved palace of the Qing Dynasty, which carries abundant historical and cultural information. As a paragon of palaces, it has historical, artistic, cultural and social values. The Prince Kung's Palace was inhabited by He Shen, a grand secretary with great political power, Princess Hexiao, a beloved daughter of Emperor Qianlong, Prince Qing Wang Yonglin, a beloved younger brother of Emperor Jiaqing, Prince Kung Yixin, who held power during the reign of Emperor Guangxu, Puwei, who attempted to rebuild the Qing Dynasty, and famous painter Puru, who was "as famous as" Zhang Daqian. Currently, the Prince Kung's Palace is an important heritage site under state protection, National 5A tourist attraction, national intangible cultural heritage display and conservation center, and national first-class museum. As a living cultural space, traditional culture and art display platform and cultural heritage place, it is used for sightseeing, museum show and intangible cultural heritage display and conservation.

1. Conservation Principles

1982 saw the enforcement of the Law of the People's Republic of China on the Protection of Cultural Relics, which strengthens the preservation of cultural relics. So far, the Law of the People's Republic of China on the Protection of Cultural Relics has been revised five times, gradually perfecting the system of law on the protection of cultural relics. The buildings in the Prince Kung's Palace are protected according to the Law of the People's Republic of China on the Protection of Cultural Relics, and all of their historical information is preserved in accordance with the principles of truthfulness and completeness. 2002 saw the promulgation of the Principles for the Conservation of Heritage Sites in China, which is a professional document used to deepen. Chinese culture heritage conservation should be promoted based on China's long-term experience in conservation of cultural relics with the Venice Charter as reference.

As an international principle of conservation of historic architecture and sites, the Venice Charter emphasizes the conservation of reveal historic architecture and the rehabilitation of historic architecture by scientific and technological means. It stipulates that pictures, carvings and ornaments shall not be rebuilt, demolished or altered, and can only be taken down for the sake of conservation, to reproduce the aesthetic value of historic architecture. The Law of the People's Republic of China on the Protection of Cultural Relics underlines the principle that the original state of historic architecture shall not be changed while it is renovated, maintained, relocated or used.

The Prince Kung's Palace has a high historical and aesthetic value, so it should be renovated and protected in strict accordance with the legal regulations on historic building conservation. Since the Prince Kung's Palace was inhabited by many owners, the historical and cultural information of every historical period is reflected in the architectural ornaments. Although it is mainly renovated based on the historical information of the reign of Emperor Tongzhi and Emperor Guangxu, the earlier ornaments are preserved as well in order to show the decorative characteristics of different periods.

"The buildings should be renovated according to the historical foundation if there is a historical foundation for them; or renovated based on expert opinions if there is no historical foundation; or renovated according to specific circumstances if there is no historical foundation or expert opinion". The above is the principle of renovating the Prince Kung's Palace, and it is the same as the principle of conservation of historic architecture in the Venice Charter and the

艺为主，现代技术为辅助手段进行修复。

恭王府的保护包含对所蕴涵的珍贵历史文化信息和营造技艺的保护，涵盖建筑、装饰、色彩、形制等级、风俗文化、营造技艺等内容，使建筑更具有生命力。周围环境也与历史街区风貌环境相协调，避免采用现代风格和过于华丽的建筑，使恭王府与周边环境形成相互联系的整体。建筑、装饰、周边环境都应得到整体性有效的保护，使其成为一个有机整体，完整综合地展现历史的风貌。

2. 技艺传承

建筑文化的价值主体存在于营造技艺和精神文化两方面。营造技艺以人为载体，是活态的遗产。长期以来各界对于非遗传承人群重视不足，导致年轻一代在主流价值观的影响下不愿从事传统营造行业，而老一代的工匠、传承人逐渐老去，营造技艺的传承没有新的活力注入而渐失光彩。传统的技艺传承方式主要是师徒传承，传承方式单一，在多元化社会中显示出局限性。

2003年联合国教科文组织在巴黎通过的《保护非物质文化遗产公约》中提到，保护人类非物质文化遗产是普遍的意愿和共同关心的事项。针对非遗传承问题，我国在2011年通过并实施《中华人民共和国非物质文化遗产法》，其中就非遗调查、非遗代表性项目名录、非遗的传承与传播等进行了规范。2015年由文化和旅游部、教育部主办的"中国非物质文化遗产传承人群研修研习培训计划"，目的在于扩宽非遗传承人眼界，提高他们的理论知识与传承水平，建立起科学有效的非遗传承机制。2016年开始，由文化和旅游部支持，相继在多地设立传统工艺工作站，尊重与发掘传统文化，扩大工艺传承队伍，发展具有地域特色的工艺品牌。政策法规的制定和非遗活动的举办，都说明我国对非遗的重视程度增加，注重对技艺和精神文化方面的传承。

对恭王府建筑装饰活态的传承，需要做好对相关技艺传承人的记录和访问。建立、记录优秀的传统工匠档案和传承谱系，提高传承人的社会地位，实现可持续的传承。结合新媒体宣传王府营造技艺与精神文化，在高校普及传统文化的知识，转变新时代下年轻人对于传统文化的认识，令更多具备深厚知识底蕴的优秀青年积极投入到未来传统建筑及非遗保护工作中，为传统建筑文化传承注入新的活力。

3. 文化传播

王府文化体现皇权之下，庶民之上的精神文化。涵盖了政治、经济、文化、艺术、信仰、

Law of the People's Republic of China on the Protection of Cultural Relics. The existing structure members should be retained and used as much as possible to show the historical look of the palaces of the Qing Dynasty. The stage of unavailable structure members should remain unchanged to have the historical information preserved. For the traditional decoration components, they should be processed by traditional technology with modern technology as a supplementary means.

Conservation of the Prince Kung's Palace involves the protection of the valuable historical and cultural information and construction techniques contained, including buildings, ornaments, colors, architectural form, customs and culture and construction techniques. The buildings are there given more vitality. The surroundings are in tune with the palace, so a modern style and excessively gorgeous style cannot be adopted, in order that the palace and the surroundings should form an interconnected whole. The buildings, decorations and surroundings should be protected effectively as an organic whole, to have the historical style and features revealed completely and comprehensively.

2. Inheritance of Craftsmanship

The value subject of architectural culture resides in construction technique and spiritual culture, of which the former, with man as the carrier, is a living heritage. For a long time, there has been a lack of enough attention to intangible cultural heritage inheritors, disinclining young people to work on traditional construction under the influence of mainstream values. However, the older-generation craftsmen and inheritors have grown old, gradually tarnishing the inhabitation of the traditional construction techniques, into which no new vitality is injected. The traditional techniques are primarily inherited under the apprenticeship system, and this single mode of inheritance has shown limitations in the multicultural society.

As is mentioned in the Convention for the Safeguarding of Intangible Cultural Heritage issued by the UNESCO in Paris in 2003, protecting the intangible cultural heritages of humanity is what most people are willing to do and concerned about. To solve the issue of intangible cultural heritage inheritance, China passed and enforced Intangible Cultural Heritage Law of the People's Republic of China in 2011. In the law there are provisions on surveys of intangible cultural heritages, lists of representative intangible cultural heritages and inheritance and transmission of intangible cultural heritages. In 2015, the Ministry of Culture and Tourism and Ministry of Education jointly initiated "Program for Advanced Training of Inheritors of Chinese Intangible Cultural Heritages" for the aim of broaden inheritors' vision and improve their theoretical knowledge and inheritance ability to build a scientific and effective intangible cultural heritage inheritance mechanism. In 2016, traditional craftsmanship centers began to be set up one after the other in many places with support from the Ministry of Culture and Tourism, in a bid to respect and excavate traditional culture, expand the craftsmanship inheritance team, and develop craftsmanship brands with regional characteristics. The enactment of policies and regulations and holding of intangible cultural heritage events reveal China's increasing attention to intangible cultural heritages, as well as the inheritance of craftsmanship and spiritual culture.

To inherit the living architecture decorations in the Prince Kung's Palace, we need to record and interview related craftsmanship inheritors. A record and pedigree excellent traditional artisans can help to enhance inheritors' social status, to achieve sustainable inheritance. We can publicize the palace's construction craftsmanship

宗教等多方面内容。恭王府建筑群是展示清王府文化的重要载体。自 2008 年完全对公众开放后，游客逐年增多，来自各地的游客络绎不绝来一睹昔日和珅、奕訢住所的风采，感受浓厚的文化氛围。2012 年荣膺国家 AAAAA 级景区，2017 年获批国家一级博物馆。

恭王府建筑群是历史的印记，反映着清王府文化。对恭王府文化传播方式的探索和解读，为其它王府未来的发展和王府文化的传播有重要的参考价值。王府文化的传播途径主要有举办展览展演展研活动、复原场景、开发文创产品、新媒体等方式。

（1）举办展览、展示、展演活动。

举办"三展"活动可以吸引游客，通过举办一些王府文化和非遗相关的精品展览，一方面，游客通过展览在恭王府建筑空间中欣赏建筑装饰，加深对于恭王府文化艺术的认识。另一方面，增加游客的文化体验，令王府文化得到更好得传播。恭王府府邸目前常设展览有 9 个，对王府历史、宗教、文物、陈设等方面进行了展示（表 1）。恭王府福文化展以"福"

and spiritual culture, and popularize traditional cultural knowledge in universities to change contemporary young people's knowledge of the traditional culture, so that more excellent young people with profound knowledge would throw themselves into the conservation of traditional buildings and intangible cultural heritages to inject new vitality into the inheritance of the traditional architectural culture.

3. Culture Communication

The prince's palace culture reflects a spiritual culture below the imperial power and above the common people. It contains aspects of contents such as politics, economy, culture, art, belief and religion. The building complex in the Prince Kung's Palace is an important carrier for display of the prince's palace culture of the Qing Dynasty. Since it was fully opened to the public in 2008, the number of visitors has increased year by year. Visitors come from all over the world in an endless stream for a glimpse of the palace that was once inhabited by He Shen and Yixin, to feel the dense cultural atmosphere. In 2012, the palace was named a national 5A scenic spot. In 2017, it was named a national first-class museum.

The building complex in the Prince Kung's Palace is a historical imprint, which reflects the prince's palace culture of the Qing Dynasty. An exploration and interpretation of the mode of the transmission of the Prince Kung's Palace culture is of great reference value to the future development of the palace and the transmission of its culture. The prince's palace culture is mainly transmitted by holding exhibitions, restoring scenes, developing cultural creative products, and propagandizing it on new media.

表 1 恭王府府邸常设展览
Tab. 1 Long-term Exhibitions in the Prince Kung's Palace

展览名称 Exhibition Tile	展览地点 Exhibition Venue
清代王府文化展 Exhibition on the Prince's Palace Culture of the Qing Dynasty	银安殿 Yin'an Palace
"守护者——北京海关 30 年罚没文物调拨恭王府特展" "Guardian - Special Exhibition on the Cultural Relics confiscated by Beijing Customs in the Past 30 Years and transferred to the Prince Kung's Palace"	银安殿东西配殿 East and West Side Halls in Yin'an Palace
水法楼原状陈列 Display of Shuifa Pavilion in Original State	水法楼 Shuifa Pavilion
多福轩室内复原陈列展 Display of Recuperative Furnishings in Duofu Hall	多福轩 Duofu Hall
锡晋斋明清家具特展 Special Exhibition on Ming and Qing Furniture in Xijin Hall	锡晋斋 Xijin Hall
恭王府历史沿革展 Exhibition on the History of the Prince Kung's Palace	葆光室 Baoguang Room
紫玉金砂——恭王府馆藏紫砂作品展 Purple Jade & Gold Sand - Exhibition on the Purple Clay Ware kept in the Prince Kung's Palace	后罩楼福善缘展厅 Fusshanyuan Showroom in Houzhao Building
恭王府福文化展 Exhibition on the Blessing Culture in the Prince Kung's Palace	后罩楼东侧福文化展厅 Blessing Culture Showroom in Houzhao Building
神音禅韵——恭王府的宗教生活展 Divine Voice & Zen Charm - Exhibition on the Religious Life in the Prince Kung's Palace	后罩楼无量居 Wuliang Room in Houzhao Building

文化为主题，展示"福"文化的历史脉络，令观众感受长久以来人们对福祉的精神追求。多福轩室内复原陈列展是按照奕訢时期的室内陈设原貌进行复原展览，展示了奕訢会客的场景，向观众传达清代亲王会客礼仪与历史。流失文物是体现传统文化的重要物件，"守护者——北京海关30年罚没文物调拨恭王府特展"展示了流失文物。临时展览涉及书画、非遗、摄影等多方面精品展览。

恭王府是国家非物质文化遗产展示与保护中心。展演活动以恭王府为平台，在2011年与中国昆剧古琴研究会合作，于每年的文化遗产日举办"良辰美景·恭王府非遗演出季"活动，邀请久负盛名的昆剧、古琴艺术大家和代表性的国家级非遗传承人在恭王府大戏楼进行展演，呈现出最富昆曲规范性和创造力的艺术作品，重释经典的意义，在2017年非遗演出季上，首次结合VR技术和互联网直播手段，是文化遗产日期间的一个重要文化品牌。"锦绣中华"主题展系列活动从2016年开始，在恭王府举办了两场展览展演。2016年"锦绣中华——传统刺绣技艺精品展"、2017年"锦绣中华——中国非物质文化遗产服饰秀"系列活动——苏绣精品展。两场活动以静态展览、服饰秀展演、"活态"传承人现场展示来诠释优秀传统文化的创造性转化和创新性发展。

除了展览展演，还开展学术研讨活动。2016年伴随着"锦绣中华——传统刺绣技艺精品展"，还举办"传承与传播——非遗传承人时尚设计师"学术研讨会。2017年在"锦绣中华——中国非物质文化遗产服饰秀"系列活动之后举办传统服装服饰与纺织印染绣技艺传承与保护的学术研讨会，围绕苏绣项目"传承与振兴"的主题，展开了深入交流与讨论。这些活动的开展推动非遗保护事业与当代经济社会的融合与发展，令优秀传统艺术重现魅力和光芒。

展览、展示、展演、展研活动相互结合，交相呼应，充分发挥了博物馆的文化传播职能，增加游客的文化体验，对拓展王府文化的研究领域，丰富清王府文化内涵有着重要意义。

（2）复原场景。

室内装饰与陈设是王府文化的重要载体。复原场景是将恭王府建筑的室内情况根据历史文献、老照片等史料复原文物和室内原貌，展示生活场景。例如多福轩室内复原陈列、后罩楼佛楼等，为观众呈现生动的历史原貌与风采。

多福轩在恭亲王奕訢时期，是奕訢的会客厅，在功能上还兼具书房的功用。2008年恭王

(1) Holding Exhibitions.

Visitors can be attracted to the palace by holding top-quality "exhibitions" on the palace culture and intangible cultural heritages. On the one hand, while visiting architectural decorations in the palace, visitors can deepen their awareness of the palace culture and art. On the other hand, the palace culture can be better transmitted after visitors gain stronger culture experience. In the Prince Kung's Palace there are 9 long-term exhibitions, which involve the history of the palace, religion, cultural relics and furnishings in the palace (Tab.1). The Prince Kung's Palace Blessing Culture Exhibition, themed on "blessing", aims to display the historical context of "blessing" culture, to show visitors how people have been pursuing happiness spiritually since long ago. The furnishings in the Duofu Hall are restored according to the mode of furnishings when the palace was inhabited by Yixin displaying how Yixin receives a guest, showing visitors the prince's hospitality. Looted relics are important articles about the traditional culture. Looted relics are displayed at "Guardian - Special Exhibition on the Cultural Relics confiscated by Beijing Customs in the Past 30 Years and transferred to the Prince Kung's Palace". Temporary exhibitions are held to display painting and calligraphy, intangible cultural heritages and photos.

The Prince Kung's Palace is a national intangible cultural heritage display and conservation center. Exhibitions and performances are given in the Prince Kung's Palace. In 2011, the palace developed cooperation with China Kun Opera & Guqin Research Institute to hold "Moments of Beauty—Prince Kung's Palace Intangible Cultural Heritage Performance Season" on the Cultural Heritage Day each year. Prestigious Kunqu Opera artists and guqin performers and representative national intangible cultural heritage inheritors are invited to the Grand Opera Tower in the Prince Kung's Palace to perform typical Kunqu Opera programs and creative artworks to reinterpret the meaning of classics. The event is an important culture brand concerning the Cultural Heritage Day. In the 2017 season, VR technology and live webcast were employed for the first time. "Splendid China" Exhibition came into being in 2016 and was held twice in the Prince Kung's Palace. In 2016, "Splendid China—Exhibition on Traditional Top-quality Embroidery Techniques" was held. In 2017, "Splendid China —Chinese Intangible Cultural Heritage Fashion Show" series - Exhibition on Top-quality Suzhou Embroidery was held. During the above two events, static exhibitions, fashion shows and "living" inheritor shows were held to explain the creative transformation and innovative development of excellent traditional culture.

Apart from exhibitions and performances, academic discussions are made as well. In 2016, during the "Splendid China - Exhibition on Traditional Top-quality Embroidery Techniques", an academic conference on "Inheritance & Transmission - Intangible Cultural Heritage Inheritor Fashion Designer" was held. In 2017, following the "Splendid China—Chinese Intangible Cultural Heritage Fashion Show" series, an academic conference on the inheritance and protection of traditional clothing, spinning and weaving, printing and dyeing techniques was held, and a deep discussion was made on the inheritance and revitalization" of Suzhou embroidery. These events are carried out in order to promote the integrative development of the intangible cultural heritage protection cause and contemporary economy and society, to reproduce the charm of excellent traditional arts.

While exhibitions, displays, performances and researches are combined together, the museum becomes the best able to transmit

府管理中心在收集和考证大量史料的基础上，依据历史资料、研究成果、现状条件复原多福轩室内空间。复原后的多福轩正殿明敞，屏门横楣上方为"尚德延釐"匾额，两侧抱柱有楹联，屏门前是地平书案，在屏门两侧是通壁书架，东西板壁为四建柜和条案。在英法联军入侵北京后，奕訢与英法侵略者谈判地点就在多福轩。复原后的多福轩充分再现了恭亲王奕訢在多福轩会客的场景与活动。

另一方面，随着国家文化的发展，影视城、文化小镇、文创园等相继开发，重塑传统四合院，将其打造成为观光、度假、购物、会议等多功能园区成为一种潮流。例如在珠海文创园区的平沙电影小镇中，将重塑恭王府，结合南、北方特色建成建筑群，既可以发展影视拍摄、旅游观光、休闲娱乐等事业，又向大众传播王府文化。

（3）文创产品。

恭王府丰厚的文化资源有助于文创产品的开发。文创产品将建筑元素作为设计语言，通过概况、抽象、结合多种艺术风格等方法运用在合适的产品中。以现代设计的角度，赋予产品优秀传统文化内涵，令产品兼备实用功能并满足人们精神层面需求。在文创产品的销售与宣传方面，将产品分类整合，采取线上、线下销售模式。通过微博、微信、APP等新媒体平台，结合恭王府的历史文化进行宣传与推广，吸引观众及消费者的注意。人们购买文创产品可以令融合在文创产品中恭王府建筑装饰文化价值得到广泛地传播。

（4）新媒体宣传。

新媒体的发展成为传播信息的主要途径。微博、直播平台、移动终端等多元化媒体方式改变了传统的文化传播方式。结合跨界的思维方式，以互联网络为平台，在传承和传播之间建立桥梁，传播王府文化。

4. 小结

恭王府文化空间探索，从历史、文化、社会的视角总结了文化艺术价值体系。随着国家对于非物质文化遗产的重视程度增加，除了对清王府建筑本体的保护，更多的关注于非物质文化遗产的传承，列举了这两个层面的保护传承的方式。清王府建筑是官式建筑的重要组成部分，体现着北京古都历史文化。恭王府作为国家一级博物馆，承担着文化传播的职责。在多元化背景下，提出了文化传播的方式，既可以完整地继承历史文化信息，又能在一定程度上丰富博物馆服务和旅游文化。习总书记指出"传承中华文化，绝不是简单复古，也不是盲

culture and enhance visitors' cultural experience. This is of great significance to expand the scope of research on the prince's palace culture and enrich the connotation of the prince's palace culture.

(2) Scene Restoration.

Interior decorations and furnishings are an important carrier of the Prince's Palace culture. Scene restoration means restoring the original appearance of the interiors of the Prince's Palace and related cultural relics according to historical documents and old photos. For instance, the original appearance of the furnishings in Duofu Hall and Buddhist Tower in Houzhao Building can be restored to show visitors the historic original appearance.

When the palace was inhabited by Prince Kung Yixin, Duofu Hall was a reception room, where he also read books. In 2008, on the basis of collecting and researching a large amount of historical data, the Prince Kung's Palace Management Center restored the interior space of Duofu Hall according to historical materials, research results and the conditions of the hall. After restoration, the main hall in Duofu Hall becomes bright and spacious. Above the lintel of the screen door is a horizontal inscribed board that says "Upholding Virtue and Integrity". On the columns on both sides there are couplets; in front of the screen door is a console writing desk; on both sides of the screen door are wall-mounted bookshelves; on the east and west wooden partitions are a cabinet and long narrow table. After the Anglo-French Allied Forces invaded Beijing, Yixin negotiated with the invaders in Duofu Hall. The restoration fully reproduces the scenes that Prince Kung Yixin received guests in Duofu Hall.

On the other hand, along with the development of the national culture, film and TV cities, cultural towns and cultural creative parks have been developed one after another, making it a trend to rebuild the traditional courtyard dwelling house by building it into a multifunctional park that serves as a sightseeing destination, resort, shopping center and meeting room. For example, in Pingsha Film Town in Zhuhai Cultural Creative Park, the Prince Kung's Palace will be renovated and buildings with north and south Chinese characteristics will be built to develop businesses, such as film shooting, sightseeing and amusement, and transmit the prince's palace culture to the public.

(3) Cultural Creative Products.

The abundant cultural resources in the Prince Kung's Palace help to develop cultural creative products. With architectural elements as a design language, multiple artistic styles can be summarized, abstracted and combined together to make cultural creative products. The products can be endowed with excellent traditional culture connotations from perspective of modern design, to be practical and able to meet people's spiritual demands. In terms of marketing and promotion, the cultural creative products can be classified and integrated together to be sold online and offline. The products can be promoted and marketed on new media such as microblog, WeChat and APP based on the history and culture of the Prince Kung's Palace, to attract attention from visitors. The people that buy the cultural creative products can widely propagate the cultural value of the architectural decorations contained in the products.

(4) New Media Propaganda.

New media has become a main way by which information is disseminated. Diverse modes of medium, such as microblog, live broadcast platform and mobile terminal, have changed the traditional mode of cultural transmission. The Internet can be used a platform in multiple modes of thinking to transmit the prince's place culture by building a bridge between inheritance and transmission.

目排外，而是古为今用、洋为中用，辩证取舍、推陈出新，拼弃消极因素、继承积极思想，以先人之规矩，开自己之生面，实现中华文化的创造性转化和创新性发展"。传统建筑装饰需要结合现代的方式进行保护传承和弘扬清王府官式建筑装饰艺术，传播王府文化。对清王府建筑装饰艺术文化还需继续深入研究，继续探索文化传承的模式，铸就中华优秀传统文化新辉煌。

作者单位：北京建筑大学建筑与城市规划学院

4.Summary

Based on the above exploration on the cultural space in the Prince Kung's Palace, a cultural and artistic value system is established from a historical, cultural and social perspective. As China pays increasing attention to intangible cultural heritages, more attention has been paid to the inheritance of intangible cultural heritages, in addition to the protection of the Prince's Palace Architecture of the Qing Dynasty. The modes of protection and inheritance of these two aspects are enumerated. the Prince's Palace Architecture, which is an important part of official architecture, embodies the history and culture of Beijing as an old capital. As a national first-class museum, the Prince Kung's Palace is obligated to spread culture. Under the background of multiplicity, a mode of cultural transmission is put forward. It can not only help to completely inherit the historical and cultural information, but also enrich museum services and tourism culture to some extent. General Secretary Xi says, "To inherit the Chinese culture, we cannot restore ancient ways or reject foreign things blindly, but should adapt ancient forms for present-day use, adapt foreign things for Chinese use, make our choice dialectically, bring forth the new through the old, abandon negative factors, inherit positive thinking, and open up a fresh outlook for ourselves with the ancestors' established practice, so as to promote the creative transformation and innovative development of the Chinese culture." Traditional architectural decorations need to be protected and inherited in a modern way. Moreover, we should carry forward the official architectural decorative art of the Prince's Palace of the Qing Dynasty and spread the palace culture. Also, we should make more intensive studies on architectural decorative art of the Prince's Palace, and go on exploring a way to inherit culture, to once again make excellent traditional Chinese culture brilliant.

多元联合教学模式的探讨——由"室内设计6+"联合毕业设计教学活动引发的思考

An Exploration on Multi-Joint Teaching——Reflections on "Interior Design 6+" Joint Graduation Project Event

王一涵

Wang Yihan

"室内设计6+"联合毕业设计起始于2013年，是国内多所建筑、艺术院校与企业联合举办的教学活动，迄今已是连续举办第六届。而本科毕业设计是高等院校本科人才培养的核心内容，也是实现本科培养目标的重要教学手段。联合设计的教学模式为毕业设计的局限性打开了一个多元融合共生的教育平台。各校师生能够在这个平台上，以教学的方式共同探讨当下社会变革中的典型案例和焦点议题，交流教学理念和方法，增进学术理论研究和实践经验的分享，从而形成一种多学校、跨专业、多地域、多语言、多思维、多元文化的教学模式。

1. "传统"与"联合"

建筑学、环境设计、风景园林等设计类专业有明显的学科特征，学生们既要掌握系统全面的专业与理论知识，又要与不断变革中的社会接轨。因此，联合毕业设计将多学科、多学校校企合作的教学模式以及教学活动与具体项目联系起来，真题真做，使学生在实践中发现和处理实际问题。"多学科交叉式的教学模式"以及"国际联合设计的教学模式"早在国外已经出现很多年。这种多元联合教学模式的目的是为培养具备创新能力以及综合素质的复合型人才。

多元联合的教学模式有效的解决了传统毕业设计中存在的缺陷与弊端。

首先，传统毕业设计的教学模式多为一名老师指导数名学生，思维的交流便局限于二者之间。联合设计则是毕设教学在多所联合学校平行展开，期间设有场地调研，中期交流和期终评图三次集中教学活动。其中参与点评的专家不仅有各大高校的专家、教授，还有知名企业的资深设计师。学生能够从中汲取不同人的设计观点，拓宽设计的创作与创新意识。

其次，传统毕业设计指导没有团队合作意识，多数为一个人独立完成一个题目，缺少了相互配合，集体完成任务的能力。而联合设计的教学模式要求组建学生设计团队，以小组的形式进行项目设计与汇报。任务书解读、现场调研、问题归纳分析以及专项设计等环节比较适合以小组的形式完成。合作过程中往往会激发出令人意想不到的成果。学生是多专业联合毕业设计的重要组成部分，基于项目模式的毕业设计，按照项目的设计内容来组建跨专业的学生设计团队。例如：在本次联合毕业设计中，我校将建筑工程学院建筑专业与艺术学院环艺设计专业的学生两两组合，根据项目的工作量制定小组人数，并推选出团队负责人，负责整个项目的协调工作。在毕业设计的教学中我们应该多鼓励学生进行团队合作，为将来的学术科研和设计生产打下一定的实践基

The "Interior Design 6+" Joint Graduation Project Event, founded in 2013, is a teaching activity jointly carried out by several Chinese universities of architecture and art in cooperation with enterprises. So far, it has been held for 6 years. Undergraduate graduation project is the core content of undergraduate talent training at institutions of higher learning, and an important teaching means by which undergraduate training objectives are achieved. As a teaching mode, joint design solves the limitations of graduation project by offering a diverse educational platform. On the platform, teachers and students at the universities involved can discuss typical cases and focus issues about present social changes, exchanges ideas with one another concerning teaching ideas and methods, and enhance the sharing of academic theoretical research and practical experience, to form a interschool, trans-disciplinary, interregional, multi-language, multi-thinking and multicultural teaching mode.

1. "Traditional" and "Joint"

The design specialty, including architecture, environmental design and landscape architecture, has obvious disciplinary characteristics, so students should not only systematically acquire professional and theoretical knowledge, but also bring themselves in line with the changing society. Therefore, the joint graduation project event connects multi-disciplinary, interschool and school-enterprise cooperative teaching and teaching activities with specific projects, to help students identify and solve practical problems in practice. "Interdisciplinary teaching" and "international joint design teaching" have come into being for many years abroad. This diverse teaching mode aims to produce interdisciplinary talents with innovative abilities and comprehensive qualities.

Multi-joint teaching has effectively solved the defects of traditional gradation design.

First, in traditional graduation project teaching, a teacher has to guide many students, so only two people can exchange ideas with each other. However, joint design enables graduation project teaching to be carried out among several universities through three intensive teaching activities: field research, in-process exchange and final comments. Moreover, the commenters are experts or professors from the universities or senior designers from famous enterprises. Students can learn design from different people, so as to broaden their design creation and innovative consciousness.

Second, a traditional graduation project team does not have teamwork awareness, and individual students often need to complete a topic alone, so they are incapable of working together and fulfilling a task together. However, joint design teaching requires setting up a student design team for the sake of project design and presentation. The team members can work together to interpret the assignment, make a field research and sort and analyze problems. In the process of cooperation, they may make unexpected achievements. Students are an important component of multidisciplinary joint graduation project, so a professional student design team needs to be established according to project design content based on the project mode concerned. For instance, in the current joint graduation project event n, our university has two teams, one of which is made up of the students majoring in architecture from School of Architectural Engineering, the other of the students majoring in environmental art design from School of Art. The number of team members is decided based on workload, and a team leader is appointed to be responsible for project coordination. In graduation project teaching, we should frequently encourage students to do the teamwork, to lay a practical

础，特别是在多所院校参与的联合教学过程中，跨高校和跨专业之间的合作交流更是一次难得的优势互补机会。

最后，传统毕业设计的题目多为导师模拟命题，缺乏创新性，有些毕业设计的题目甚至是虚拟的，缺少实际工程背景，存在很大的局限性。毕业设计的选题及任务书的制定是整个教学体系中最基础和最关键的环节，它直接决定了后续教学活动的开展。设计选题和任务书的制定按年度由参与高校根据自身的地域特色和专业背景轮流提出备选方案，最终由多所院校联合商讨制定，并且在多个方面体现出行业特色。联合毕业设计的教学模式则每届的选题都会针对社会热点问题，由承办高校当地的设计单位出题，实事实做，十分具有挑战性。例如："室内设计 6+"联合毕业设计就是参加高校分别联合相关企业，形成"7所参与高校 -7 家合作企业 -7 类设计课题"的活动新格局。从第一届的北京鸟巢国家体育场赛后改造设计，到上海地铁（交通大学站）改造环境设计，从南京晨光 1865 创意产业园环境设计，到北京密云耀阳国际老年公寓环境改造设计，再到传统民居保护性利用设计，题目针对性强，具有鲜明的时代感，给毕业生同学提供了一个很好的设计实践机会。同时院校背景及地域属性的不同必然导致了专业认知的差异，而这种差异正是联合教学活动中最具特色的支撑点。

2. 多元化复合型人才的培养

由此可见，多元联合教学模式应以学科特性为前提，以交叉融合为目的，通过多元化的教学模式来提高学生系统性、整体化的思维模式，增强学生在实践过程中协同合作的团队意识，以及各项专业技能的规范性与综合性的完善。

这种联合教学模式首先培养的是"复合型"人才的综合性能。在联合设计中，通过与不同学校的学生、导师、不同企业的设计师相互交流，不同专业的设计者相互配合，从而将自我的单一知识向复合知识进行转变，使知识的维度从一维向多维扩展。从学科特性出发，设计从历史文脉、政策经济、城市结构、地域特性、建筑更新、景观环境等多维角度进行思考。从整个设计的全过程来提高学生发现问题、分析问题、最终解决问题的综合性与系统性的能力。

其次，联合设计是培养学生跨学科协作的整体思维与概念，即"大建筑"概念。也就是说无论是规划、建筑、景观、环艺的学生，都要具备一个综合所有学科的整合意识。从而全面掌握从实地调研、方案创作、技术应用、规范制图的设计标准。同时要鼓励全面的实践性教学，因此校企联合就提供了一个极好的实践平台。企业的参

foundation for future academic research and design. Especially, in the process of joint teaching attended by several universities, inter-university and trans-disciplinary cooperation and exchanges can help well complement each other's advantages.

Finally, most traditional graduation project topics are assigned by supervisors, lack of innovativeness. Even some graduation project topics are virtual ones, lacking practical engineering background, so they have great limitations. For graduation project, topic selection and assignment compilation are the most basic and important part of the whole teaching system, and directly decide the entire teaching activity. For topic selection and assignment compilation, the universities involved take turns to put forward alternatives according to their own regional characteristics and professional backgrounds. Finally, they hold a discussion to finalize a topic and assignment and embody industry characteristics in many aspects. For joint graduation project teaching, every year's topic is a social hotspot, which is finalized by the university itself with the design company on a realistic basis, so the topic is high-challenge. Taking the "Interior Design 6+" Joint Graduation Project Event for example, it is attended by "7 universities and 7 enterprises with 7 design topics". From the post-game transformation design of Beijing Bird's Nest National Stadium in the first year to the environmental reconstruction design of Shanghai Metro (Jiao Tong University Station); from the environmental design of Nanjing Chenguang 1865 Creative Industry Park to the environmental renovation design of Beijing Yaoyang International Apartments for the Elderly and to the protective utilization design of traditional dwellings, every year's topic is issue-relevant and has distinctive characteristics of the times. Moreover, the difference in background and region leads to a difference in professional cognition, and the difference is the most distinctive supporting point in joint teaching.

2. Interdisciplinary Talent Training

Thus it can be seen that multi-joint teaching should be subject to disciplinary features, and aim to fuse different subjects together, to improve students' systematic and holistic thinking mode, enhance their teamwork awareness in practice, and improve the normativity and comprehensiveness of various professional skills.

This mode of joint teaching first fosters "interdisciplinary" talents' comprehensive qualities. In joint design, my knowledge is extended with mono-disciplinary knowledge turning into multi-disciplinary knowledge after communicating and working with the students and teachers from different universities and designers from different enterprises. From the perspective of disciplinary features, we should think about design in many aspects including historic content, policy and economy, urban structure, regional characteristics, building renewal and landscape environment. In the entire design process, the students can become more capable of identifying, analyzing and solving problems.

Second, joint design develops students' holistic thinking ability in interdisciplinary collaboration. In other words, they think under the idea of "grand architecture". That is to say, all students, including everyone majoring in planning, architecture, landscape or environmental art, should consciously combine all disciplines together, so as to acquire proficiency in the design standards concerning field research, scheme creation, technique application and standard charting. Besides, we should advocate comprehensive practical teaching, so university-enterprise cooperation provides a very good practical platform. Enterprise's participation gives topic source social,

与使项目的课题来源有了社会性、事实性与实践性，企业参观实习、实地案例的考察以及企业导师参与毕设指导的各个环节，都有利于学生提高设计的实际性，更好的与即将毕业后走向的工作岗位接轨。

最后，联合设计能够培养学生整体最优化的设计评价方法。设计是一门综合性极强的学科，整体最优的评价方法是指综合城市、区域、经济、文化、生态、技术等综合性指标，从而构建一种整体的认知观，在专业分工与综合设计之间培养独立工作与相互协作的融合能力。在寻找整体最优化的前提下，使个体服从整体，使团队中的学生能够构建一种在设计过程中综合分析、判断与决策的设计评价系统。

3. 教学成果的交流与共享

联合毕业设计教学成果的交流展示与共享是整个联合教学环节中最为重要的组成部分。在这一环节中，不仅学生能够通过答辩、展览，传递自身的创作，同时也能从其他学校的学生作品中受到启发。这种分享与交流不仅对学生有着重要的指导意义，同样也为高校教师与企业专家进行专业交流和教学方式方法的探讨提供了极具价值的研究意义。例如此次"室内设计6+"联合毕业设计教学活动参与评价和答辩的人员广泛涵盖不同高校和学科的成员，这样可使评价的标准更为多元、学生接触的知识更为全面。集中的评价和答辩不少于三次，哈尔滨工业大学开题汇报、南京艺术学院中期汇报以及同济大学答辩汇报，多次集中评价分别针对学生的现状分析及不同阶段的成果进行科学的、有针对性的指导。同时结合具体案例和专项设计展开集中讲座、论坛等活动。最终的成果形成联合设计作品集、图纸展板等；同时联合中国水利水电出版社正式出版发行等多种方式来完成设计成果的交流与共享。

4. 结语

校际间与校企联合展开的多元毕业设计教学模式，是各大高校顺应社会与时代背景以及专业发展中探索互补与开放的"多元化复合型"人才培养的新模式。这种模式改变了现有的毕业设计教学模式与教学方法，有利于共享校内外的各类教学资源，加强学生的综合实践能力锻炼。对学生的专业知识培养、教师教学方法更新、高校教学模式改革都具有重大的价值和意义。多元联合教学模式能够积极整合教学资源和科研成果，从而形成专业合作与校际、校企之间的交流，提供共享平台。这种教学模式不但有利于优化人才的培养模式，也更能使行业向着多元化健康发展。

作者单位：浙江工业大学建筑工程学院

factual and practical significance. Visit and practice in the enterprise, on-the-spot investigation in it, and the corporate designer's guidance on graduation project help to enhance the practicality of students' design, so that they could better bring themselves in line with future operating posts.

Finally, joint design can help students to grasp a globally optimum design evaluation method. Design is a highly comprehensive discipline. A globally optimum design evaluation method refers to building a holistic cognitive perspective by combining urban, regional, economic, cultural, ecological and technological indexes together to foster independent-working and mutual-collaborating fusion abilities between professional division of labor and comprehensive design. On the premise of seeking optimization of whole, let individuals obey the whole, and enable the team members to build a design evaluation system that can be used for comprehensive analysis, judgment and decision-making in the process of design.

3. Sharing of Teaching Achievements

Display and sharing of joint graduation project teaching achievements is the most important part of the whole joint graduation project teaching process. In this part, students can not only share with each other their works through oral defense and display, but also inspire one another. Not only is such sharing of great guiding significance to students, but it is of valuable significance to university teachers and enterprise experts' professional communication with each other and exploration on teaching methods. For instance, the "Interior Design 6+" Joint Graduation Project Event is attended by many university teachers and students. In this way, evaluation criteria are diversified and students can get more comprehensive knowledge. Evaluation and oral defense are carried out collectively for at least times, including Harbin Institute of Technology's opening report, Nanjing University of the Arts' in-process report and Tongji University's oral defense report. Multiple joint evaluations are made in order to analyze the students' current situation and offer scientific and targeted guidance on the achievements made at different stages. Meanwhile, lectures and forms are jointly carried out for specific cases and particular design. The final achievements are compiled into a joint design work portfolio and displayed on boards; meanwhile, the achievements are published by China WaterPower Press for further display and sharing.

4. Conclusions

The inter-university university-enterprise joint graduation project event is a new complementary and open mode of "interdisciplinary talent" training carried out by the universities in order to conform to the development of the society, times and specialties. This mode changes the existing graduation project teaching mode and method, and helps to share various intramural and extramural teaching resources and enhance students' comprehensive practical abilities. This is of great value and significance to help students acquire professional knowledge, help teachers renew teaching methods and help universities reform teaching modes. Multi-joint teaching can help to actively integrate teaching resources and scientific research achievements together to provide a shared platform for professional cooperation, inter-university exchanges and university-enterprise exchanges. This teaching mode can only optimize personnel training modes, but also enable the industry to achieve diversified healthy development.

热点命题，纷呈特色

联合指导、服务需求

专家讲坛

老建筑保护与重生
Conservation and Renascence of Old Buildings

宋微建

上海微建（Vjian）建筑空间设计有限公司 董事长

余秋雨先生说："废墟是文化的使节"，老建筑是历史文化的载体，也是研究历史实物的例证，更是一座城市的记忆，是城市历史的见证者，它承载着这座城市的文化积淀。

关于老建筑的修复与保护，有不同声音。对此问题，欧洲分成了三大流派，法国（19世纪）的风格派主张老建筑"像它应该的那样，甚至比原来更好"认为老建筑要结合当代社会环境，有自己的风格；英国（19世纪）的废墟派则主张老建筑"富有诗意的死亡"，"修古建，古建亡"认为老建筑不应修复更不应重建；意大利的历史派主张"要保护，不要恢复"，"禁止装饰性复原"认为老建筑要适当的修复，但不能改变其本质。

1964年《威尼斯宪章》是国际上第一部保护和修复文物建筑的法规，是历史文化遗产保护发展中一个重要的里程碑。其中的四项原则：原真性、整体性、可辩识性、可逆性，是我们修复老建筑的原则参照。1994年的《奈良真实性文件》

Mr. Yu Qiuyu says, "Ruins are an envoy of culture." An old building is the carrier of historical culture, a historical object, and the memory of a city. It is a witness of city history, and carries the culture of the city.

As for the renovation and conservation of old buildings, there are different points of view. Behind this issue are three schools in Europe. The De Stijl, a French school (of the 19th century), which advocates "keeping an old building like what it should be or making it better", argues that an old building should be renovated based on the contemporary social environment, with its own style unchanged; the Ruins School, a British school (of the 19th century), which advocates "perishing an old building poetically" and thinks that "an old building will perish if it is renovated", argues that an old building shouldn't be renovated or rebuilt; the Italian historical school, which advocates "protecting old building rather than renovating it", argues that an old building should be renovated properly with its essence unchanged.

The Venice Charter, enacted in 1964, is the world's first rule on the conservation and renovation of historic buildings, and an important milestone in the conservation and development of historical and cultural heritages. There are four principles in it: authenticity, integrity, identifiability and reversibility,

重新定义了遗产保护界最基本的概念——真实性，把这个原本有严格定义和科学评判标准的概念加以"多样化"，各文化均可自行定义其"真实性"的内涵。有关"真实性"的事例，不得不提"掘立柱"和"式年造替"。"掘立柱"最大的缺点就是柱子的寿命问题。长年埋于地下，难免要受害虫的侵袭。因此有了"式年造替"制度。"式年造替"指的是两块比邻的用地，一地营造一地休闲。每隔20年在基地旁边把原有建筑物依照原样新建一次。休闲地只保留有砂地，种植树木，去除围合。日本伊势神宫至今还保留着"式年造替"的传统，内宫、外宫都保有两块相同大小的社殿用地，每隔20年依古法在另一块用地重建社殿并迁祭。

老建筑保护从来都是一个难题。"中国的600多座城市都经历了旧城改造和重建，结果是每个城市的样貌几乎都变成了一个样，每一分钟，都有文化遗产在消失。再不保护，五千年历史文明古国就没有东西留存了。"产生这些问题的原因在于对文化的历史价值的忽视，与经济利益的冲突。专业人士不专业，职业操守也不够。当然，值得肯定的是，我国对文物的认识是渐进的，从文物到历史建筑，再到历史街区，文物建筑保护的范围扩大与法规条例的不断完善，让我们看到了老建筑新的希望。但立法仍不及破坏的速度，最大的危害是"保护"性的破坏。所以中国对老建筑的修护与保护依然任重道远。

which guide the renovation of old buildings. The Nara Document on Authenticity, issued in 1994, redefines the most basic concept of heritage preservation—authenticity, which is originally a well-defined and scientifically-judged concept that is, however, is "diversified" in the document, so that the connotation of "authenticity" can be defined by each culture. Among the cases of "authenticity", there are two typical modes of construction: "construction with columns buried in earth" and "reconstruction every twenty years". Of the two modes, the former has an obvious disadvantage that columns have a short life: it is inevitably damaged by worms in earth. Therefore, the latter came into being. It means that there are two adjoining land parcels, one used with the other leaving unused. The building is newly built on the idle land every 20 years. On the unused land there is only sand, but trees are planted there without enclosure. The Ise Grand Shrine is still rebuilt this way today. There are two pieces of land of the same size in the interior space and exterior space, and every 20 years, the building is newly built on the other piece of land.

阿什河·左岸 1905
Ashihe River - Left Bank 1905

韩冠恒

中国建筑学会室内设计分会理事室内设计师
2008 法国 CNAM 国立工业学院研究生毕业
现任哈尔滨唯美源装饰设计有限公司合伙人

这段曾经沉寂百年的历史；

这段中国工业制糖的先河；

这段往昔繁华的工业景象；

集中体现了中西方人民智慧的结晶，我们所希望的是使其重新产生社会价值……

无论结果怎样，作为设计师我们都珍惜参与思考的过程，认真对待这份社会责任，对自己、对历史负责……

园区以工业文化核心区为主要区域敷设和推动以动态为主的生活文化商业区，以静态为主的北方民俗体验区，烘托以创业为主文化创意产业基地，以及以体验为主文化艺术体验区，可以说工业文化核心区把五个大区紧紧的串联到一起，让五个区块相互交融、相互提升，进而形成立体式交流的创意园区。五大园区：工业文化核心区、生活文化商业区、北方民俗体验区、文化创意产业基地、未来工业艺术体验区。

This phase of history that was becalmed for a century;

This precedent of Chinese industrial sugaring;

This industrial scene that was bustling in the past;

Jointly reflect the crystallization of Chinese and Western people's wisdom, and all that we want is to make them generate social values again…

No matter what results we have, as a designer, we should cherish the process of thinking, take our social responsibility seriously, and be responsible for ourselves and history…

In the park, there is the industrial cultural core area as a primary zone, as well as a dynamic business area of life and culture, a static experience area of north Chinese folk customs, and an entrepreneurship-oriented industrial base for cultural creativity. The industrial cultural core area holds the five areas together, as it were, making them fuse with each other and promote each other, forming a three-dimensional creative park. The five areas: Industrial Cultural Core Area, Business Area of Life and Culture, Experience Area of North Chinese Folk Customs, Industrial Base for Cultural Creativity, and Experience Area of Future Industrial Art.

工业遗产保护与展示
Protection and Display of Industrial Heritage

朱海玄
哈尔滨工业大学建筑学院　副教授

18世纪末源于英国的工业革命,彻底改变了人类的技术、生活和环境,使人类发展由农业社会全面进入了工业社会。同时,英国就此走向了世界强国的道路,早在1918年,为保存谢菲尔德曾经辉煌的工业历史信息,谢菲尔德大学就成立了谢菲尔德技术贸易协会(STTS),开展工业历史与工业技术研究。1955年,伯明翰大学的Michael Rix发表《Industrial Archaeology》文章,呼吁对英国工业革命以来的工业遗产进行系统研W究和全面保护,标志着工业考古学科的正式产生。工业考古学作为一门对工业遗存进行系统的调查、记录与研究的学科,有力地支持了英国工业遗产的保护与展示研究的科学性。Blaenavon工业遗址,2000年被列入世界文化遗产名录,其工业遗存——铁矿场、采煤场、采石场、有轨交通线、炼铁炉、工人社区、公共设施等,向人们展示了19世纪南威尔士作为世界上最主要钢铁和煤炭生产基地的辉煌历史,其基于工业考古学的工业遗产保护与展示模式,具有广泛的借鉴意义。

The Industrial Revolution, which first took place in the UK in the late 18th century, thoroughly changed the human technology, life and environment, transforming the agricultural society into the industrial society. Meanwhile, the UK gradually became a world power. Back in 1918, the University of Sheffield founded the Sheffield Trades Technical Society (STTS) and carried out research on industrial history and technology in order to preserve the glorious industrial historical information of Sheffield. In 1955, Michael Rix, a professor at the University of Birmingham, published an essay in Industrial Archaeology to appeal to people to systemically research and completely protect all the industrial heritages since the Industrial Revolution. This represents the formal birth of industrial archaeology. As a subject that systemically researches and record industrial heritages, industrial archaeology offers energetic scientific support to the conservation, display and research of the British industrial heritages. Blaenavon Industrial Heritage, which was put on the World Heritage List in 2000, contains iron mines, coal pits, stone pits, transit lines, iron-fining furnaces, a work community and public facilities. The industrial heritage shows people the brilliant history of South Wales as the world's main steel and coal production base in the 19th century. It is of far-reaching referential significance that Blaenavon is protected and displayed as an industrial heritage.

热点命题，纷显特色

联合指导，服务需求

风采定格

室内设计6+2018（第六届）
联合毕业设计
Interior Design 6+ 2018 (Sixth Year)
University Participate Cooperative Graduation Project Event

"6+" 2018（第六届）联合毕业设计开题

「室内设计 6+」2018（第六届）联合毕业设计
"Interior Design 6+"2018(Sixth Year) Joint Graduation Project Event

國粹賡續卷——歷史建築保護與再利用設計

風采定格

「室内设计 6+」2018（第六届）联合毕业设计
"Interior Design 6+"2018(Sixth Year) Joint Graduation Project Event

風采定格

國粹賡續卷——歷史建築保護與再利用設計

"室内设计 6+" 2018（第六

IID-ASC
中国建筑学会室内设计分会

中国建筑学会室内设计分会（简称 IID-ASC），前身是中国室内建筑师学会，成立于1989年，是在住房和城乡建设部中国建筑学会直接领导下、民政部注册登记的社团组织，是获得国际室内设计组织认可的中国室内设计师的学术团体。

分会的宗旨是团结全国室内设计师，提高中国室内设计的理论与实践水平，探索具有中国特色的室内设计道路，发挥室内设计师的社会作用，维护室内设计师的权益，发展与世界各国同行间的合作。

中国建筑学会室内设计分会每年举办丰富多彩的学术交流活动，为设计师提供交流和学习的场所，同时也为设计师提供丰富的设计信息，促进中国室内设计行业更好更快地发展。

室内分会秘书处设在北京，负责学会相关工作。

Institute of Interior Design-ASC

Six colleges display differently and this reflects a certain difference between them.The predecessor of Institute of Interior Design-ASC (IID-ASC) is China Institute of Interior Architects. Since itsestablishment in 1989, IID-ASC has been the only authorized academic institution in the field of interior design in China.

IID-ASC aims to unite interior architects of the whole country, raise the theoretical and practical level of China's interior design industry, pioneer the Chinese characteristics of interior design, help interior architects play their social role, preserve the rights and interests of interior architects and foster professional exchanges and cooperation with international peers.

IID-ASC hold abundant and colorful academic exchanges every year, building aplatform for designers to communicate and to study meanwhile update designer information of design industry to enhance the better and rapid development of interior design industry of China.

IID-ASC Secretarial is located in Beijing, taking charge of institute work.

同济大学

同济大学创建于 1907 年,教育部直属重点大学。同济大学 1952 年在国家院系调整过程中成立建筑系,1986 年发展为建筑与城市规划学院,下设建筑系、城市规划系和景观学系,专业设置涵盖城市规划、建筑设计、景观设计、历史建筑保护、室内设计等广泛领域。同济大学建筑与城市规划学院是中国大陆同类院校中专业设置齐全、本科生招生规模最大,世界上同类院校中研究生培养规模第一,具有全球性影响力的建筑规划设计教学和科研机构,是重要的国际学术中心之一。

同济大学室内设计教育起源于建筑系,同济大学建筑系于 20 世纪 50 年代就开始注重建筑内部空间的研究,1959 年曾尝试在建筑学专业中申请设立"室内装饰与家具专门化"。1986 年经国家建设部和教育部批准,同济大学建筑系成立了室内设计专业,1987 年正式招生,成为中国大陆最早在工科类(综合类)高等院校中设立的室内设计专业。1996 年原上海建材学院室内设计与装饰专业并入同济大学建筑系;2000 年原上海铁道大学装饰艺术专业并入同济大学建筑系。2009 年同济大学开始恢复建筑学专业(室内设计方向)的招生工作。2011 年建筑学一级学科目录下,设立"室内设计"二级学科。

同济大学建筑城规学院的教学理念为以现代建筑的理性精神为灵魂,以自主创造、博采众长的学术品格为本色,以当代技术与地域文化的并重交融为导向,以国际学科前沿的跟踪交流为背景。室内设计教学突出建筑类院校室内设计教学特色,强调理性精神,提出"以人为本、关注生态、注重环境整体观、时代性和地域性并重、融科学性和艺术性于一体"的室内设计观。

Tongji University

Tongji University, established in 1907, is a top university of China Ministry of Education. During the time of restructuring of the university and college systems in 1952, the Department of Architecture was formed at Tongji University, and in 1986 was renamed as the College of Architecture Urban Planning (CAUP). Currently CAUP has three departments: the Department of Architecture, the Departmentof Urban Planning, the Department of Landscape Design. The undergraduate program covers: Architecture, Urban Planning, Landscape Design, Historic Building Protection and Interior Design. CAUP is one of China's most influential educational institutions with the most extensive programs among its peers, and the largest body of postgraduate students in the world. Today, CAUP hasbeen recognized as an international academic center with a global influence in the academic fields.

Tongji University's interior design education originated from the Department of Architecture which started to conduct interior space research in the 1950's. In 1959, it applied for the establishment of the "Interior Decoration and Furniture Specialty" within Architecture Discipline. In 1986, approved by the Ministry of Education and the Ministry of Construction, the "Interior Design Discipline" was formally founded. Starting to admit undergraduate studentsin 1987, Tongji University was one of two earliesthigh education institutions in mainland China to train interior design professionals in a University of science and technology. In 2011.

"Interior Design" officially became the secondary discipline of the Architecture Discipline. In the same year, the "Interior Design Research Team" was established, providing even broader room for subject development. Tongji University's interior design education crystallizes its own characteristics, emphasizing rational thinking and proposing the interior design concept of "human centric, ecological consciousness, overall environmental perspective, equal time and regional characteristic significance, technology and art integration".

华南理工大学

　　华南理工大学位于广东省广州市，创建于1934年，是历史悠久、享有盛誉的中国著名高等学府。是中华人民共和国教育部直属的全国重点大学、首批国家"211工程""985工程"重点建设院校之一。

　　华南理工大学设计学院组建成立于2010年6月，现有工业设计、环境设计、信息与交互设计、服装与服饰设计等4个系。设计学院紧密依托华南理工大学雄厚的理工优势和深厚的人文底蕴，积极探寻与产业高度结合和国际化合作的道路，旨在打造享誉国内外设计创新人才培养和设计实践与服务的研究高地。

　　当前，设计学院紧紧把握设计创意产业的发展契机，不断创新 教育理念，大胆探索设计创新人才培养模式，以"技术创新引领、文化创意引领、产业转型引领、可持续发展引领"为建设目标，拥有"创意与可持续设计研究院"以及"当代艺术空间""设计实验与实践公共平台""跨学科拔尖创新人才培养试验区"和"腾龙研发中心""文化艺术与创意产业研究中心""中国民间艺术研究中心""陶瓷文化研究所"等一系列产学研平台，力争建设成为国内领先、有国际影响力的设计学院，从而支撑、引领国家和广东设计产业发展。

South China University of Technology

　　South China University of Technology(SCUT), located in Guangzhou city, Guangdong province, was founded in 1934. It is a well-known Chinese university which has a long history and enjoys a high reputation. It is a national key university directly under the Ministry of Education of the People's Republic of China, one of the first national "211 project" and "985 project" key construction of colleges and universities.

　　The design institute of SCUT was founded in June 2010, with majors including industrial design, environmental design, information and interaction design, clothing and apparel design. The design institute closely relies on the strong advantage of technology and deep cultural heritage of SCUT and actively explores the way of highly industry integration and international cooperation, aiming to create famous heights for domestic and foreign design innovative talent training as well as design practice and service.

　　At present, the design institute grasps the development opportunity in design creativity industry, constantly renews education idea and makes bold exploration in design innovation personnel training mode. With "leading technology innovation, leading culture innovation, leading industry transformation and sustainable development" as the construction goal, a series of production and research platform including "creative and sustainable design and research institute" "space of contemporary art" "public platform of design experiment and practice" "interdisciplinary top creative talents cultivation test area" and "Tenglong research&development center" "cultural art and creative industry research center" "Chinese folk art research center" "ceramic culture research institute" has been built, striving to become the domestic leading design instutite with international influence, so as to support and lead the design industry development in Guangdong and across the country.

哈尔滨工业大学

哈尔滨工业大学隶属于国家工业和信息化部，是首批进入国家"211工程""985工程"和首批启动协同创新"2011计划"建设的国家重点大学。1920年，中东铁路管理局为培养工程技术人员创办了哈尔滨中俄工业学校——即哈尔滨工业大学的前身，学校成为中国近代培养工业技术人才的摇篮。学校已经发展成为一所特色鲜明、实力雄厚，居于国内一流水平，在国际上有较大影响的多学科、开放式、研究型的国家重点大学。

哈尔滨工业大学建筑学学科是我国最早建立的建筑学科之一，历经90余载风雨砥砺。建筑学院建筑学科现有建筑学、城乡规划、风景园林、环境设计4个本科专业和建筑学、城乡规划学、风景园林学3个一级学科点和设计艺术学二级学科硕士点。已获得建筑学、城乡规划学和风景园林学一级学科博士、硕士授予权，以及设计学二级学科硕士授予权，还设有建筑学一级学科博士后科研流动站。建筑学院始终秉持严谨治学、精于耕耘的文化精神，打造了一支朴实敬业、有特色、有能力、肯奉献的优秀教师团队。在本科教学、研究生培养及科学研究方面，特色鲜明，成绩显著。在寒地公共建筑设计、地域建筑设计、寒地建筑技术、建筑历史与理论、寒地城市规划与城市设计、寒地环境艺术设计等诸多方向上，均形成自己的学术特色。

Harbin Institute of Technology

Harbin Institute of Technology affiliates to the Ministry of Industry and Information Technology, and is among the first group of the national key universities to enter the national"211Project""985 Project" and to start the collaborative innovation "2011 plan". In order to train engineers, the Mid east railway authority founded the Harbin Sino Russian school in 1920, the predecessor of Harbin Institute of technology, which becomes the cradle of China's modern industry and technical personnel.The School has evolved into a distinctive, powerful, first class national key university, which is multidisciplinary, open, researchful and with international influence.

The discipline of Architecture in Harbin Institute of technology is one of the earliest architectural subjects in China, with more than 90 years'ups and downs. The school of Architecture has 4 undergraduate disciplines, including Architecture, Urban and Rural Planning, Landscape Architecture,Environmental Design, and 3 first-level disciplines, including Architecture, Urban and Rural Planning,Landscape Architecture, and secondary master's disciplines in Design and Arts. We have the first-level doctorate and master's authorization in Architecture, Urban and Rural Planning and Landscape Architecture, and secondary-discipline master's authorization in Design, and Post-doctoral Research Institute on architectural first-level discipline.With the cultural spirits of rigor and diligence,The school of Architecture has created a devoted, distinctive, qualified and dedicated teachers' team.We have gained distinctive and outstanding achievements in undergraduate teaching, postgraduate education and scientific research, and have formed our own academic characteristics in the Design of Public Buildings in Cold Region, Regional Architecture, Building Technology in Cold Region, Architectural History and Theory, Urban Planning and Designing in Cold Region and Environmental Design in Cold Region.

西安建筑科技大学

　　西安建筑科技大学坐落在历史文化名城西安,学校总占地4300余亩,校园环境优美,办学氛围浓郁。学校办学历史源远流长,其办学历史最早可追溯到始建于1895年的北洋大学,积淀了我国近代高等教育史上最早的一批土木、建筑、环境类学科精华。1956年,时名西安建筑工程学院。1959年和1963年,曾先后易名为西安冶金学院、西安冶金建筑学院。1994年3月8日,经国家教委批准,更名为西安建筑科技大学,是公认的中国最具影响力的土木建筑类院校之一及原冶金部重点大学。

　　西安建筑科技大学是以土木、建筑、环境、材料学科为特色,工程学科为主体,兼有文、理、经、管、艺、法等学科的多科性大学。学校现有 16 个院(系),其60 个本科专业面向全国第一批招生,有权招收保送生,实行本硕连读。艺术设计本科专业为陕西省特色专业。

　　西安建筑科技大学艺术学院成立于2002年4月,是由建筑学院的艺术设计专业和摄影专业本科生、机电工程学院工业设计专业本科生和新成立的雕塑专业及各专业关教师组建而成。学院现有艺术设计、工业设计、摄影、雕塑、会展艺术与技术 5 个本科专业,在校本科生1200余人。艺术设计专业被评为"国家级特色专业""省级名牌专业"。学院集聚了包括建筑、规划、景观等在内的多学科的研究人才,学科团队长期致力于西部地区地域文化研究,承担了多项国家、省部级基金课题。艺术学院积极主办(承办)国家级学术、学科建设会议;邀请国际、国内知名教授来我校进行学术交流;制定管理办法,并设立专项基金,鼓励青年教师和优秀博士生开展学术交流、国际(内)合作研究,与欧洲、亚洲地区的多所大学建立了友好合作关系。

　　学院以学生全面发展为培养目标,注重学生综合素质提高,依托各类学生组织载体和平台,开展形式多样的课外活动。注重加强学术交流与互动,邀请学者、专家和社知名人士来我院举办讲座和专题报告,开阔学生视野,改善学生知识结构,培养学生的科技、人文精神。组织学生积极参与学科竞赛,

Xi'an University of Architecture and Technology

　　Located in the historical and cultural city Xi'an, covering an area of 4300 acres, Xi'an University of Architecture and Technology has beautiful campus environment and academic atmosphere. This university has quite a longhistory, which can be dated back to the Northern University, founded in 1895. Since then, in the higher education history of modern China, this university has been accumulating the first batch of disciplines essence in civil engineering, construction and environmental class. In 1956, this university was named as Xi'an Institute of Architectural Engineering. In 1959 and 1963, it was renamed as Xi'an Institute of Metallurgy and Xi'an Institute of Metallurgy and Construction. On March 8, 1994,approved by the State Board of Education, it was renamed as Xi'an University of Architecture and Technology and was recognized as one of China's most influential civil engineering colleges and the key university of the former Ministry of colleges and the key university of the former Ministry of Metallurgical.

　　Featured by civil engineering, construction,environment and materials science, engineering disciplines as the main body, Xi'an University of Architecture and Technology is a multidisciplinary university also with liberal arts, science, economics, management, arts, law and other disciplines. The university has 16 departments, 60 undergraduate programs so it can launch the first batch of undergraduate enrollment. It also has the right to recruit students by recommendation and the right of implementation of Accelerated Degree. Undergraduate art and design program is the featured major in Shaanxi Province.

　　Founded in April, 2002, Xi'an University of Architecture and Technology was established by the undergraduates from the major of art design and photography and from mechanical and electrical engineering industrial design and the relevant teachers from newly established sculpture and other specialties. The current undergraduate majors in this college include art and design, industrial design, photography, sculpture, exhibition art and technology, with more than 1,200 undergraduate students. Art Design was named "national characteristic specialty""provincial famous professional". This university has gathered many multidisciplinary researchers,including architecture, planning, landscape, etc. All these research teams have a long history of working towards the research of western region cultures, through undertaking many national and provincial funds subjects. The Arts College has actively organized (or as the contractor) the national academic, discipline-building meetings; inviting international and domestic famous professors to come for academic exchanges. It also has developed management approach, and set up a special fund to encourage young teachers and outstanding doctoral students to carry out academic exchanges and international (inside) collaborative researches. In the meantime it has established friendly and cooperative relations with the universities in Europe, Asia and other countries.

　　The university has taken the overall development of students as its

指导、鼓励学生从事科研活动，在国内刊物上发表各类论文。学院调动教研室、资料室、实验室，多方互动，通力合作，构建了教学、科研、学生三位一体的开放性实验（工作）平台。学院培养的学生深受用人单位欢迎，毕业生供不应求。

training objectives, the improvement of theoverall quality of them as the aim to focus on. Relying on various student organizations carrier and platforms,the university has carried out various forms of extracurricular activities. And also it has focused on strengthening academic exchanges and interaction, inviting scholars, experts and celebrities to come to listen to the lectures and presentations,which can broaden the students' horizons, improve their knowledge structure and culture their spirits of science, technology and humanities. In other ways, the university organized the students to actively participate in academic competitions, and guided or encouraged students to engage in research activities, and many students have published various papers in the national magazines. The college has transferred departments, libraries, laboratories and paid multi-interactive efforts or work together to build a teaching-research-student trinity open experiment (work) platform. The graduates trained by the college have been welcomed by employers and the graduates are in short supply.

北京建筑大学

北京建筑大学是北京市和住房城乡建设部共建高校、教育部"卓越工程师教育培养计划"试点高校和北京市党的建设和思想政治工作先进高校,是一所具有鲜明建筑特色、以工为主的多科性大学,是"北京城市规划、建设、管理的人才培养基地和科技服务基地""北京应对气候变化研究和人才培养基地"和"国家建筑遗产保护研究和人才培养基地",是北京地区唯一一所建筑类高等学校。

学校源于1907年清政府成立的京师初等工业学堂。学校1977年恢复本科招生,1982年被确定为国家首批学士学位授予高校,1986年获准为硕士学位授予单位。2011年被确定为教育部"卓越工程师教育培养计划"试点高校。2012年"建筑遗产保护理论与技术"获批服务国家特殊需求博士人才培养项目,成为博士人才培养单位。2014年获批设立"建筑学"博士后科研流动站。2015年10月北京市人民政府和住房城乡建设部签署共建协议,学校正式进入省部共建高校行列。2016年5月,学校"未来城市设计高精尖创新中心"获批"北京高等学校高精尖创新中心"。2017年获批推荐优秀应届本科毕业生免试攻读研究生资格。2018年5月,获批博士学位授予单位,建筑学、土木工程获批博士学位授权一级学科点。

学校有西城和大兴两个校区。目前,学校正按照"大兴校区建成高质量本科人才培养基地,西城校区建成高水平人才培养和科技创新成果转化协同创新基地"的"两高"发展布局目标加快推进两校区建设。与住建部共建中国建筑图书馆,是全国建筑类图书种类最为齐全的高校。

学校坚持立德树人,培育精英良才。现有各类在校生11842人,已形成从本科生、硕士生到博士生和博士后,从全日制到成人教育、留学生教育全方位、多层次的办学格局和教育体系。多年来,学校为国家培养了6万多名优秀毕业生,他们参与了北京60年来重大城市建设工程,成为国家和首都城市建设系统的骨干力量。学校毕业生全员就业率多年来一直保持在95%以上,2014年进入"全国高校就业50强"行列。

学校面向国际,办学形式多样。学校始

Beijing University of Civil Engineering and Architecture

Beijing University of Civil Engineering and Architecture is a university co-constructed by Beijing City and the Ministry of Housing and Urban-Rural Development, a pilot university of the "Excellent Engineer Training Program" initiated by the Ministry of Education, a university active in the Party building and ideological and political work of Beijing City, an engineering-based multiversity with outstanding architectural features, a "base for training of planning, construction and management personnel and a high-tech service base" in Beijing, a "Climate change treatment research institute and personnel training base" in Beijing, a "national architectural heritage conservation and research and personnel training base", and the only university of architecture in Beijing.

The university was formerly known as Beijing Primary Technical School, which was founded in 1907. The university resumed undergraduate admissions in 1977, was identified as one of the first undergraduate universities in 1982, authorized to award master's degrees in 1986, and identified as a pilot university of the "Excellent Engineer Training Program" initiated by the Ministry of Education in 2011. In 2012, its "architectural heritage conservation theory and technology" was approved as a national special doctoral talent training program, making it a doctoral talent training institution. In 2014, the university built a center for post-doctoral studies on "architecture". In October 2015, the Beijing Municipal Government and Ministry of Housing and Urban-Rural Development signed a co-construction agreement, listing the university among the universities co-constructed by the province and ministry. In May 2016, the university's "Sophisticated Innovation Center for Future City Design" was renamed "Beijing Sophisticated Innovation Center for Institutions of Higher Education". In 2017, the university became eligible to recommend fresh undergraduate graduates to receive postgraduate studies without sitting for the entrance examinations. In May 2018, the university was approved as a doctor's degree granter, and its architecture and civil engineering were upgraded to first-level disciplines for doctor's degree.

The university has two campuses, which are located in Xicheng District and Daxing District respectively. Currently, the university is accelerating the construction of the two campuses to build the "campus in Daxing District into a base for high-quality undergraduate talent training and the campus in Xicheng District into a base for high-level talent training and transformation of technological innovation achievements as well as collaborative innovations". The Chinese Architectural Library in it, built with the Ministry of Housing and Urban-Rural Development, is a library with the fullest range of architectural books in China.

The university insists on educating students by virtue and raising elites. Now the university has 11842 students, including undergraduate, postgraduate, doctoral and post-doctoral students, and a multi-level educational system covering full-time teaching, adult education and international students education. Over the years, the university has

终坚持开放办学战略，广泛开展国际教育交流与合作。目前已与美国、法国、英国、德国等28个国家和地区的63所大学建立了校际交流与合作关系。

站在新的历史起点上，学校将以党的十九大精神为指引，深入学习贯彻习近平新时代中国特色社会主义思想，按照"提质、转型、升级"的工作方针，围绕立德树人的根本任务，全面推进内涵建设，全面深化综合改革，全面实施依法治校，全面加强党的建设，持续增强学校的办学实力、核心竞争力和社会影响力，以首善标准推动学校各项事业上层次、上水平，向着把学校建设成为国内一流、国际知名、具有鲜明建筑特色的高水平、开放式、创新型大学的宏伟目标奋进。

provided the society with more than 60000 excellent graduates, who have participated in the major urban construction projects in Beijing in the past 60 years, becoming the backbone of the national and Beijing's urban construction systems. Its graduate employment rate has been above 95% for many times, and ranked among "China's top 50" in 2014.

The university offers forms education in the face of the world. It always insists on open schooling, and conducts international educational exchanges and cooperation extensively. Currently, the university has built an intercollegiate exchange and cooperation relationship with 63 universities in 28 countries and regions such as America, France, Britain and Germany.

Standing on a new historical starting point, the university will profoundly learn and implement Xi Jinping's thought on socialism with Chinese characteristics for a new era under the guidance of the spirit of the 19th CPC National Congress, comprehensively promote the connotation construction, fully deepen the comprehensive reform, fully implement schooling by law, fully enhance the Party building, constantly strengthen its schooling strength, core competitiveness and social influence around the fundamental task of educating students by virtue under the direction of the work policy of "quality improvement, transformation and upgrading", and increase the level of its various undertakings by considering itself the greatest philanthropist, in a bid to make itself a domestic first-class and internationally known high-level, open innovative university with outstanding architectural features.

南京艺术学院

南京艺术学院是我国独立建制创办最早并延续至今的高等艺术学府。下设 14 个二级学院，27 个本科专业及 50 个专业方向。拥有艺术学学科门类下设的艺术学理论、音乐与舞蹈学、戏剧与影视学、美术学以及设计学全部 5 个一级学科的博士、硕士学位授予权及博士后科研流动站。

南京艺术学院从 2005 年开设了展示设计本科专业和硕士专业研究方向；2008 年该专业并入工业设计学院，2011 年会展艺术与技术专业作为独立的二级学科获得国家教育部的正式批准，2012 年该专业又被归为设计学类，成为"艺术与科技"专业。南京艺术学院工业设计学院的艺术与科技（展示设计）专业以学生为中心，以学术为导向，以实践为手段，以发展为目标，通过近 10 年的发展，已经逐步形成知识融贯、结构合理、连贯而开放的模块化专业课程体系和走向现代化、全球化的课程内容。旨在为文化部门、博物馆部门、大中型展馆、设计团体、旅游部门、会展机构等单位培养具有一定的理论素养，专业知识合理，专业特点突出，具备问题导入、市场导入和文化导入的整合设计和研究能力，以及高度艺术造型及表达能力的专业设计人才。

Nanjing University of the Arts

Nanjing University of the Arts is one of the earliest arts institutions in China. It consists of 14 schools, 27 undergraduate majors and 50 major directions. It has Master's and Doctoral degrees and Post-doctoral stations in 5 subdisciplines under the first-class discipline of thearts: Arts Theory, Music and Dance, Drama and Film, Fine Arts Theory and Design Theory.

The professional background: in 2005, Display Design was set up as a major direction in undergraduate level and a research direction in master program in Nanjing University of the Arts; in 2008, it was incorporated into industrial design major as its one direction in School of Industrial Design; in 2011, it was approved by the Ministry of Education as an independent sub-discipline in then national disciplinary classification in undergraduate education; in 2012, it was classified into the first-class discipline of design with a new major name of "Art and Technology ".Through nearly 10 years of efforts by adhering to the principle that is students centered, academy-oriented, practice focused and development-guided, Art and Technology (Display Design) major has formed a coherent and open modular curriculum system of coherent knowledge and rational structure supported by modernization and globalization oriented course contents. The major is to cultivate professional design talents for the cultural sector,the museum sector, medium and large exhibition halls, the design community, the tourism sector, exhibition and other institutions. The graduates of this major are to have the ability to do design and research in the manner of integrating question, market and culture. And they are also to be cultivated as talents with the capacity of high-level artistic formation and excellent expression, as well as rational expertise structure and outstanding professional features.

浙江工业大学

浙江工业大学是一所教育部和浙江省共建的省属重点大学,其前身可以追溯到1910年创立的浙江中等工业学堂。经过几代工大人的艰苦创业和不懈奋斗,学校目前已发展成为国内有一定影响力的综合性的教学研究型大学,综合实力稳居全国高校百强行列。

2013年浙江工业大学牵头建设的长三角绿色制药协同创新中心入选国家2011计划,成为全国首批14家拥有"2011协同创新中心"之一的高校。目前学校有本科专业68个;硕士学位授权二级学科101个;博士学位授权二级学科25个;博士学位授权一级学科5个;博士后流动站4个。学科涵盖哲学、经济学、法学、教育学、文学、理学、工学、农学、医学、管理学、艺术学等11大门类。学校师资力量雄厚,拥有中国工程院院士2人、共享中国科学院和中国工程院院士3人、国家级有突出贡献中青年专家6人、国家级教学名师3人、国家杰出青年基金获得者3人、中央千人计划入选者2人、教育部长江学者特聘教授1人、教育部创新团队1个、国家级教学团队2个、各类国家级人才培养计划入选者26人次。浙江工业大学坚持厚德健行的校训,把提高教育质量放在突出位置,努力培养能够引领、推动浙江乃至全国经济和社会发展的精英人才。

Zhejiang University of Technology

Zhejiang University of Technology is a key comprehensive college of the Zhejiang Province; its predecessor can be traced back to the founding in 1910 as Zhejiang secondary industrial school.After several generations' hard working and unremitting efforts, the school now has grown to be a comprehensive University in teaching and researching which is very influential. The comprehensive strength ranks the top colleges and universities. In 2009, Zhejiang province people's government and the Ministry of Education signed a joint agreement; Zhejiang University of Technology became the province ministry co construction universities.

In 2013 Zhejiang University of Technology led the construction of Yangtze River Delta green pharmaceutical Collaborative Innovation Center which was selected for the national 2011 program,to become one of the first 14 of 2011 collaborative innovation center. There are 68 undergraduate schools; 101 grade-2 subjects of master's degree authorization; 25 grade-2 subjects of doctor's degree authorization; 4 postdoctoral research stations. Subjects include philosophy, economics, law,education, literature, science, engineering, agriculture, medicine, management, arts and other 11 categories. School teacher is strong. There are 2 Chinese academicians of Academy of Engineering,sharing 3 academicians of Chinese Academy of Sciences and Academy of Engineering; 6 national young experts with outstanding contributions; 3 National Teaching Masters, 3 winners of national outstanding youth fund, 2 people were selected to central thousand person plan, the Ministry of education,1 professor of the Yangtze River scholars, 1 innovative team of Ministry of Education, 2 national teaching teams, and 26 person were selected to all kinds of national personnel training plans.Zhejiang University of Technology adheres to its motto "Profound accomplishment and invigorating practice. Accumulate virtues and good practice." To improve the quality of education in a prominent position, and strive to cultivate to lead, promote Zhejiang and even the country's economic and social development of elite talent.

致 谢
Acknowledgements

同济大学建筑设计研究院（集团）有限公司
上海市政工程设计研究总院（集团）有限公司
广州象城建筑设计咨询有限公司
哈尔滨工业大学建筑设计研究院
荣禾集团德和建筑设计事务所有限公司
文化部恭王府中华传统技艺保护与研究中心
永隆家具
深圳世纪光华照明技术有限公司南京分公司
南京观筑历史建筑文化研究院
杭州国美建筑设计研究院有限公司

上海微建（Vjian）建筑空间设计有限公司董事长　宋微建
哈尔滨工业大学建筑学院副教授　　　　　　　　朱海玄
哈尔滨唯美源装饰设计有限公司合伙人　　　　　韩冠恒